LONDON MATHEMATICAL SOCIETY LECTURE NOTE SERIES

Managing Editor: Professor J.W.S. Cassels, Department of Pure Mathematics and Mathematical Statistics, University of Cambridge, 16 Mill Lane, Cambridge CB2 1SB, England

The titles below are available from booksellers, or, in case of difficulty, from Cambridge University Press.

34	Representation theory of Lie groups, M.F. ATIYAH *et al*
36	Homological group theory, C.T.C. WALL (ed)
39	Affine sets and affine groups, D.G. NORTHCOTT
46	p-adic analysis: a short course on recent work, N. KOBLITZ
50	Commutator calculus and groups of homotopy classes, H.J. BAUES
59	Applicable differential geometry, M. CRAMPIN & F.A.E. PIRANI
66	Several complex variables and complex manifolds II, M.J. FIELD
69	Representation theory, I.M. GELFAND *et al*
74	Symmetric designs: an algebraic approach, E.S. LANDER
76	Spectral theory of linear differential operators and comparison algebras, H.O. CORDES
77	Isolated singular points on complete intersections, E.J.N. LOOIJENGA
79	Probability, statistics and analysis, J.F.C. KINGMAN & G.E.H. REUTER (eds)
83	Homogeneous structures on Riemannian manifolds, F. TRICERRI & L. VANHECKE
86	Topological topics, I.M. JAMES (ed)
87	Surveys in set theory, A.R.D. MATHIAS (ed)
88	FPF ring theory, C. FAITH & S. PAGE
89	An F-space sampler, N.J. KALTON, N.T. PECK & J.W. ROBERTS
90	Polytopes and symmetry, S.A. ROBERTSON
91	Classgroups of group rings, M.J. TAYLOR
92	Representation of rings over skew fields, A.H. SCHOFIELD
93	Aspects of topology, I.M. JAMES & E.H. KRONHEIMER (eds)
94	Representations of general linear groups, G.D. JAMES
95	Low-dimensional topology 1982, R.A. FENN (ed)
96	Diophantine equations over function fields, R.C. MASON
97	Varieties of constructive mathematics, D.S. BRIDGES & F. RICHMAN
98	Localization in Noetherian rings, A.V. JATEGAONKAR
99	Methods of differential geometry in algebraic topology, M. KAROUBI & C. LERUSTE
100	Stopping time techniques for analysts and probabilists, L. EGGHE
101	Groups and geometry, ROGER C. LYNDON
103	Surveys in combinatorics 1985, I. ANDERSON (ed)
104	Elliptic structures on 3-manifolds, C.B. THOMAS
105	A local spectral theory for closed operators, I. ERDELYI & WANG SHENGWANG
106	Syzygies, E.G. EVANS & P. GRIFFITH
107	Compactification of Siegel moduli schemes, C-L. CHAI
108	Some topics in graph theory, H.P. YAP
109	Diophantine analysis, J. LOXTON & A. VAN DER POORTEN (eds)
110	An introduction to surreal numbers, H. GONSHOR
113	Lectures on the asymptotic theory of ideals, D. REES
114	Lectures on Bochner-Riesz means, K.M. DAVIS & Y-C. CHANG
115	An introduction to independence for analysts, H.G. DALES & W.H. WOODIN
116	Representations of algebras, P.J. WEBB (ed)
117	Homotopy theory, E. REES & J.D.S. JONES (eds)
118	Skew linear groups, M. SHIRVANI & B. WEHRFRITZ
119	Triangulated categories in the representation theory of finite-dimensional algebras, D. HAPPEL
121	Proceedings of *Groups - St Andrews 1985*, E. ROBERTSON & C. CAMPBELL (eds)
122	Non-classical continuum mechanics, R.J. KNOPS & A.A. LACEY (eds)
124	Lie groupoids and Lie algebroids in differential geometry, K. MACKENZIE
125	Commutator theory for congruence modular varieties, R. FREESE & R. MCKENZIE
126	Van der Corput's method of exponential sums, S.W. GRAHAM & G. KOLESNIK
127	New directions in dynamical systems, T.J. BEDFORD & J.W. SWIFT (eds)
128	Descriptive set theory and the structure of sets of uniqueness, A.S. KECHRIS & A. LOUVEAU
129	The subgroup structure of the finite classical groups, P.B. KLEIDMAN & M.W.LIEBECK
130	Model theory and modules, M. PREST
131	Algebraic, extremal & metric combinatorics, M-M. DEZA, P. FRANKL & I.G. ROSENBERG (eds)
132	Whitehead groups of finite groups, ROBERT OLIVER
133	Linear algebraic monoids, MOHAN S. PUTCHA
134	Number theory and dynamical systems, M. DODSON & J. VICKERS (eds)
135	Operator algebras and applications, 1, D. EVANS & M. TAKESAKI (eds)
136	Operator algebras and applications, 2, D. EVANS & M. TAKESAKI (eds)
137	Analysis at Urbana, I, E. BERKSON, T. PECK, & J. UHL (eds)

138 Analysis at Urbana, II, E. BERKSON, T. PECK, & J. UHL (eds)
139 Advances in homotopy theory, S. SALAMON, B. STEER & W. SUTHERLAND (eds)
140 Geometric aspects of Banach spaces, E.M. PEINADOR and A. RODES (eds)
141 Surveys in combinatorics 1989, J. SIEMONS (ed)
142 The geometry of jet bundles, D.J. SAUNDERS
143 The ergodic theory of discrete groups, PETER J. NICHOLLS
144 Introduction to uniform spaces, I.M. JAMES
145 Homological questions in local algebra, JAN R. STROOKER
146 Cohen-Macaulay modules over Cohen-Macaulay rings, Y. YOSHINO
147 Continuous and discrete modules, S.H. MOHAMED & B.J. MÜLLER
148 Helices and vector bundles, A.N. RUDAKOV et al
149 Solitons, nonlinear evolution equations and inverse scattering, M. ABLOWITZ & P. CLARKSON
150 Geometry of low-dimensional manifolds 1, S. DONALDSON & C.B. THOMAS (eds)
151 Geometry of low-dimensional manifolds 2, S. DONALDSON & C.B. THOMAS (eds)
152 Oligomorphic permutation groups, P. CAMERON
153 L-functions and arithmetic, J. COATES & M.J. TAYLOR (eds)
154 Number theory and cryptography, J. LOXTON (ed)
155 Classification theories of polarized varieties, TAKAO FUJITA
156 Twistors in mathematics and physics, T.N. BAILEY & R.J. BASTON (eds)
157 Analytic pro-p groups, J.D. DIXON, M.P.F. DU SAUTOY, A. MANN & D. SEGAL
158 Geometry of Banach spaces, P.F.X. MÜLLER & W. SCHACHERMAYER (eds)
159 Groups St Andrews 1989 volume 1, C.M. CAMPBELL & E.F. ROBERTSON (eds)
160 Groups St Andrews 1989 volume 2, C.M. CAMPBELL & E.F. ROBERTSON (eds)
161 Lectures on block theory, BURKHARD KÜLSHAMMER
162 Harmonic analysis and representation theory, A. FIGA-TALAMANCA & C. NEBBIA
163 Topics in varieties of group representations, S.M. VOVSI
164 Quasi-symmetric designs, M.S. SHRIKANDE & S.S. SANE
165 Groups, combinatorics & geometry, M.W. LIEBECK & J. SAXL (eds)
166 Surveys in combinatorics, 1991, A.D. KEEDWELL (ed)
167 Stochastic analysis, M.T. BARLOW & N.H. BINGHAM (eds)
168 Representations of algebras, H. TACHIKAWA & S. BRENNER (eds)
169 Boolean function complexity, M.S. PATERSON (ed)
170 Manifolds with singularities and the Adams-Novikov spectral sequence, B. BOTVINNIK
171 Squares, A.R. RAJWADE
172 Algebraic varieties, GEORGE R. KEMPF
173 Discrete groups and geometry, W.J. HARVEY & C. MACLACHLAN (eds)
174 Lectures on mechanics, J.E. MARSDEN
175 Adams memorial symposium on algebraic topology 1, N. RAY & G. WALKER (eds)
176 Adams memorial symposium on algebraic topology 2, N. RAY & G. WALKER (eds)
177 Applications of categories in computer science, M. FOURMAN, P. JOHNSTONE, & A.. PITTS (eds)
178 Lower K- and L-theory, A. RANICKI
179 Complex projective geometry, G. ELLINGSRUD et al
180 Lectures on ergodic theory and Pesin theory on compact manifolds, M. POLLICOTT
181 Geometric group theory I, G.A. NIBLO & M.A. ROLLER (eds)
182 Geometric group theory II, G.A. NIBLO & M.A. ROLLER (eds)
183 Shintani zeta functions, A. YUKIE
184 Arithmetical functions, W. SCHWARZ & J. SPILKER
185 Representations of solvable groups, O. MANZ & T.R. WOLF
186 Complexity: knots, colourings and counting, D.J.A. WELSH
187 Surveys in combinatorics, 1993, K. WALKER (ed)
189 Locally presentable and accessible categories, J. ADAMEK & J. ROSICKY
190 Polynomial invariants of finite groups, D.J. BENSON
191 Finite geometry and combinatorics, F. DE CLERCK et al
192 Symplectic geometry, D. SALAMON (ed)
193 Computer Algebra and Differential Equations, E. TOURNIER (ed)
196 Microlocal analysis for differential operators, A. GRIGIS & J. SJÖSTRAND
197 Two-dimensional homotopy and combinatorial group theory, C. HOG-ANGELONI,
    W. METZLER & A.J. SIERADSKI (eds)
198 The algebraic characterization of geometric 4-manifolds, J.A. HILLMAN

London Mathematical Society Lecture Note Series. 193

Computer Algebra
and Differential Equations

Edited by

E. Tournier
Université Joseph Fourier, Grenoble

CAMBRIDGE
UNIVERSITY PRESS

CAMBRIDGE UNIVERSITY PRESS
Cambridge, New York, Melbourne, Madrid, Cape Town, Singapore,
São Paulo, Delhi, Dubai, Tokyo, Mexico City

Cambridge University Press
The Edinburgh Building, Cambridge CB2 8RU, UK

Published in the United States of America by
Cambridge University Press, New York

www.cambridge.org
Information on this title: www.cambridge.org/9780521447577

First published 1994

A catalogue record for this publication is available from the British Library

Library of Congress cataloguing in publication data available

ISBN 978-0-521-44757-7 Paperback

Contents

1 Effective methods in $\mathcal{D}$-modules.   **1**

   1.1 Motivations and introduction to the theory of $\mathcal{D}$-modules. *(B. Malgrange).*   3

   1.2 $\mathcal{D}$-modules : an overview towards effectivity. *(Ph. Maisonobe).*   21

2 Theoretical aspects in dynamical systems.   **57**

   2.1 Introduction to the Ecalle theory *(E. Delabaere).*   59

   2.2 Perturbation analysis of nonlinear systems *(K. R. Meyer).*   103

3 Normal forms.   **141**

   3.1 Normal forms of differential systems *(J. Della Dora, L. Stolovitch).*   143

   3.2 Versal normal form computation and representation theory *(J. A. Sanders).*   185

   3.3 Painlevé analysis and normal forms *(L. Brenig, A. Goriely).*   211

   3.4 Normal forms and Stokes multipliers of nonlinear meromorphic differential equations *(Y. Sibuya).*   239

Acknowledgements

We would like to thank the Centre National de la Recherche Scientifique, the Université Joseph Fourier de Grenoble, and the IMAG Institute for their financial support of this workshop.
We would also like to thank Claire Di Crescenzo for her very useful help in the preparation of this book with Latex.

Evelyne Tournier

Preface

When the Organizing Commitee of CADE began to choose the program of CADE-92, it was decided that $\mathcal{D}$-modules would be a central topic at this conference.

The theory of $\mathcal{D}$-modules is quite recent. It began in the late sixties and at first was considered to be quite abstract and difficult. Over the years the situation improved with the development of the theory and its applications. The organizers felt that it was time to try to introduce it to a larger audience interested in differential equations and computer algebra, since the theory of $\mathcal{D}$-modules offers an excellent way to effectively handle linear systems of analytic PDEs.

Once this decision was made it was natural to ask Bernard MALGRANGE to be the *"invité d'honneur"* at CADE-92, with the task of lecturing about $\mathcal{D}$-modules in a way adapted to an audience interested in effectivity. This was natural because Bernard MALGRANGE is not only one of the most famous mathematicians in this field, but also because he is perhaps the true originator of this direction. It is generally admitted that $\mathcal{D}$-module theory began in the early seventies with the fundamental work of I. N. BERSTEIN and of the Japanese school around M. SATO, but in fact Bernard MALGRANGE introduced the basic concepts ten years ago for the constant coefficients case (see his 1962 Bourbaki report *"systèmes différentiels à coefficients constants"*), and later for the general case (see his lectures at Orsay *Cohomologie de Spencer (d'après Quillen)*).

I think it was quite a challenge to explain the basic ideas of $\mathcal{D}$-module theory to an audience of non-specialists with such a variety of interests, but Bernard MALGRANGE helped by A. GALLIGO and Ph. MAISONOBE took up this challenge.

I hope that the readers of the book will enjoy this introduction as much as the participants of CADE-92 did. Moreover this volume also contains many other interesting approaches to computer algebra and differential equations, such as theoretical aspects in dynamical systems and normal forms.

Jean-Pierre RAMIS

Chapter 1

Effective methods in $\mathcal{D}$-modules

B. Malgrange

Motivations and introduction to the theory of $\mathcal{D}$-modules

Ph. Maisonobe

$\mathcal{D}$-modules : an overview towards effectivity

Motivations and introduction to the theory of $\mathcal{D}$-modules

Bernard Malgrange *

Contents

1. Introduction   4

2. Finiteness properties   6

3. Dimension theory   7

4. Holonomic modules in one variable   12

5. Operations on $\mathcal{D}$-modules   17

6. References   20

*Institut Fourier, Université Joseph Fourier, BP 74, F-38402 St Martin d'Hères Cedex

1.   Introduction

Generally speaking, one describes a "$\mathcal{D}$-module" as a module over a ring of differential operators; for instance, we will work with the following ones:

– $A = \mathbf{C}[x, \partial]$, the ring of (linear) differential operators with polynomial coefficients; here $x = (x_1, \dots, x_n)$; $\partial = (\partial_1, \dots, \partial_n)$, $\partial_j = \frac{\partial}{\partial x_j}$. This ring can also be thought of as the ring of non-commutative polynomials in $2n$ variables x_i, ∂_i, with the "Heisenberg commutation relations" $[x_i, x_j] = 0$, $[\partial_i, \partial_j] = 0$, $[\partial_i, x_j] = \delta_{ij}$, the Kronecker symbol.

– $\mathcal{D} = \mathcal{O}[\partial]$, $\mathcal{O} = \mathbf{C}\{x\}$, the ring of convergent series; therefore $\mathcal{D}$ is the ring of differential operators with analytic coefficients at $0 \in \mathbf{C}^n$.

– $\hat{\mathcal{D}} = \hat{\mathcal{O}}[\partial]$, $\hat{\mathcal{O}} = \mathbf{C}[[x]]$, the ring of formal power series.

More generally, it is often useful to work with sheaves of modules over a sheaf of rings of differential operators (e.g., over an open set of $\mathbf{C}^n$, or over an analytic manifold). For simplicity, we will not consider this case here.

Take e.g. the case of $\mathcal{D}$; the others are similar. A (left) $\mathcal{D}$-module can be thought of in two different ways.

i) A space of "functions" on which the differential operators act. Here, e.g. the space of C^∞ functions near 0 in $\mathbf{R}^n$; or the space of analytic functions at 0 in $\mathbf{R}^n$ or $\mathbf{C}^n$; or the space of germs at 0 of distributions in $\mathbf{R}^n$ etc... . This is normal, and the language of $\mathcal{D}$-modules will add nothing new here.

ii) A system of linear partial differential equations. This requires some explanation; denote by $\mathcal{C}$ any one of the spaces of functions considered above. A system of equations in $\mathcal{C}$ is given by a map $A \cdot : \mathcal{C}^p \to \mathcal{C}^q$, $A = (a_{ij})$, $1 \le i \le q$, $1 \le j \le p$, $a_{ij} \in \mathcal{D}$ (explicitly, such a system is $\sum a_{ij} f_j = g_i$, f_j and $g_i \in \mathcal{C}$).

To this system, we associate a left $\mathcal{D}$-module in the following way: denote by $\cdot A$ the map $\mathcal{D}^q \to \mathcal{D}^p$ defined by $b \mapsto bA$, b being the row vector $(b_1, \dots, b_q) \in \mathcal{D}^q$. Put $M = \operatorname{coker}(\cdot A)$, i.e. by definition, $\mathcal{D}^p/\mathcal{D}^q A$; as $\cdot A$ is clearly compatible with the structure of left $\mathcal{D}$-modules of $\mathcal{D}^p$ and $\mathcal{D}^q$, M is a left $\mathcal{D}$-module (but it has no natural structure of a right $\mathcal{D}$-module).

Denote by $\ker(A \cdot)$ the "homogeneous system" associated to A, i.e. the set of $f = \begin{pmatrix} f_1 \\ \vdots \\ f_p \end{pmatrix} \in \mathcal{C}^p$ such that $Af = 0$. The next proposition shows that $\ker(A \cdot)$ "depends only on M".

Proposition 1.1. — *One has an isomorphism* $\ker(A \cdot) \simeq \operatorname{Hom}_{\mathcal{D}}(M, \mathcal{C})$.

As usual, the right-hand side denotes the set of $\mathcal{D}$-linear maps $M \to \mathcal{C}$. To prove this one applies $\mathrm{Hom}_\mathcal{D}(\cdot, \mathcal{C})$ to the (exact) sequence $\mathcal{D}^q \to \mathcal{D}^p \to M \to 0$; this gives a sequence $0 \to \mathrm{Hom}_\mathcal{D}(M, \mathcal{C}) \to \mathcal{C}^p \to \mathcal{C}^q$ (note that the map $u \mapsto u(1)$ gives an isomorphism $\mathrm{Hom}_\mathcal{D}(\mathcal{D}, \mathcal{C}) = \mathcal{C}$). The last sequence is again exact by the so-called "left exactness" of $\mathrm{Hom}_\mathcal{D}(\cdot, \mathcal{C})$. See any treatise on homological algebra; or, better, verify by yourself that it is obvious!

Remark 1.2. — By using *resolutions* of M one can also express the equations with right-hand side in terms of M. I will not do it here, since it is a little bit more complicated and I will not need it. See in any treatise on homological algebra the general nonsense on "Hom" and "Ext^i".

Now, the question is: Why replace the simple notion of differential system by the complicated notion of $\mathcal{D}$-module? Of course, this is unnecessary to study a given equation as, e.g. the Laplace equation; the reasons are different:

i) One can have different "presentations" of M, e.g. different ways of represent it as a quotient as above. This gives different systems, whose solutions correspond to each other.

Actually these systems are, in a sense, "trivially equivalent" (as occurs, e.g. when one adds to the unknown functions some of their derivatives as new unknowns, adding also of course the corresponding equations). I leave it as a good exercise to the reader to make a precise statement. Anyway, the consideration of a $\mathcal{D}$-module gives a way of reasoning independent of the special system which represents it; this is sometimes useful.

ii) More generally, we can be interested, not in a special equation, given a priori, but in the general theory, and especially in some systems or classes of systems which would not be so easy to write explicitly; one tries to prove some general properties of such systems (these properties could eventually be useful for explicit calculations).

We will see several examples of such properties in the following sections. Let me give here only one example:

Take $f \in \mathcal{O}$, $f \neq 0$, and consider the ring $\mathcal{O}[f^{-1}]$ of meromorphic functions at $0 \in \mathbf{C}^n$ of the form g/f^k, $f \in \mathcal{O}$, $k \in \mathbf{N}$. This is obviously a $\mathcal{D}$-module. One can prove that it is generated over $\mathcal{D}$ by $1/f^k$, for $k \geq k_0$ (k_0 depending on f). For such a k, the map $a \mapsto a(\frac{1}{f^k})$ represents $\mathcal{O}[f^{-1}]$ as $\mathcal{D}/\mathcal{I}_k$, $\mathcal{I}_k$ the ideal annihilating $1/f^k$. Finding the corresponding presentation means just finding a basis of $\mathcal{I}_k$, which in general is not at all obvious!

[*Incidentally*: two left ideals $\mathcal{I} \subset \mathcal{D}$ and $\mathcal{J} \in \mathcal{D}$ can look very different and nevertheless give the same module. The simplest example is the following:

take $n = 1$, $M = \mathcal{O} = \mathbf{C}\{x\}$. Then $\mathcal{O}$ is generated on $\mathcal{D}$ by 1; therefore $\mathcal{O} = \mathcal{D}/\mathcal{I}$, $\mathcal{I} = \mathrm{Ann}(1)$; it is easy to verify that $\mathcal{I}$ is generated by ∂. On the other hand, one has also $\mathcal{O} = \mathcal{D}x = \mathcal{D}/\mathcal{J}$, $\mathcal{J} = \mathrm{Ann}(x)$; it is easy to verify that $\mathcal{J}$ is not generated by one element. A minimal system of generators is, e.g. $x\partial - 1$, ∂^2.]

iii) A third, and very important, reason to consider $\mathcal{D}$-modules is the following: in this language, it is possible to define some natural *operations* e.g. direct and inverse images.

But these operations, except in some simple cases, have no obvious definition directly in terms of differential systems; and the corresponding systems are difficult to calculate explicitly.

I shall say a few words on these operations in section 5.

2.   Finiteness properties

I will keep the notations of the preceding section. I shall state the results for $\mathcal{D}$, but they are equally true for $\hat{\mathcal{D}}$ and A.

Theorem 2.1. — *The ring $\mathcal{D}$ is left and right nœtherian.*

Recall that "left nœtherian" means the following: let M be a left $\mathcal{D}$-module "finite over $\mathcal{D}$", i.e. admitting a finite system of generators; then, any $\mathcal{D}$-submodule of M is also finite.

It is sufficient to prove the result for $M = \mathcal{D}$, i.e. to prove that left ideals are finite (hint: use induction on the number of generators of M). To prove this last result, we use the *filtration* of $\mathcal{D}$ by the degree of the operators: let us say that $a = \sum a_\alpha \partial^\alpha$ $(a_\alpha \in \mathcal{O}, \partial^\alpha = \partial_1^{\alpha_1} \cdots \partial_n^{\alpha_n})$ is of degree $\leq k$ if $a_\alpha = 0$ for $|\alpha| = \alpha_1 + \cdots \alpha_n > k$; denote by $\mathcal{D}_k$ the operators of degree $\leq k$. Then the graded ring $\mathrm{gr}\, \mathcal{D} = \oplus(\mathcal{D}_k/\mathcal{D}_{k-1})$ is commutative since $[\mathcal{D}_k, \mathcal{D}_k] \subset \mathcal{D}_{k+\ell-1}$; actually it is equal to $\mathcal{O}[\xi]$, $\xi = (\xi_1, \ldots, \xi_n)$, with $\xi_i = \mathrm{gr}\, \partial_i$. Now, the result is a consequence of the following statements:

i) The ring $\mathcal{O}$ is nœtherian (classical: preparation theorem, standard basis, etc... .).

ii) Since $\mathcal{O}$ is nœtherian, $\mathcal{O}[\xi]$ is nœtherian; this is a classical result of Hilbert.

iii) Since $\mathrm{gr}\, \mathcal{D}$ is nœtherian, $\mathcal{D}$ itself is left (and right) nœtherian.

This is also classical, but I will recall the proof. Let $\mathcal{I}$ be a left ideal of $\mathcal{D}$; the ideal $\mathrm{gr}\, \mathcal{I}$ of $\mathrm{gr}\, \mathcal{D}$ is constructed with the "principal symbols" of operators of $\mathcal{I}$ (explicitly, if $a = \sum\limits_{|\alpha| \leq k} a_\alpha \partial^\alpha$, the principal symbol, more precisely the principal symbol of order k, is $\sigma(a) = \sum\limits_{|\alpha|=k} a_\alpha \xi^\alpha)$. Now, by $ii)$, $\operatorname{gr} \mathcal{I}$ has a finite basis, $(\bar{a}_1, \ldots, \bar{a}_p)$; we can suppose that a_i is homogeneous, say of degree m_i; let $a_1, \ldots, a_p \in \mathcal{D}$, of degrees respectively $m_1, \ldots, m_p$ with $\sigma(a_i) = \bar{a}_i$. We will be finished if we prove the following lemma.

Lemma 2.2. — *The a_i's generate $\mathcal{I}$ over $\mathcal{D}$.*

Let $b \in \mathcal{I}$, $\deg b \leq \ell$, and let $\sigma(b)$ be its principal symbol (of degree ℓ); we have $\sigma(b) = \sum \bar{c}_i \bar{a}_i$, $\bar{c}_i \in \operatorname{gr} \mathcal{D}$; by homogeneity, we can suppose that $\bar{c}_i$ is homogeneous, of degree $\ell - m_i$; choose $c_i \in \mathcal{D}$, $\deg c_i \leq \ell - m_i$, $\sigma(c_i) = \bar{c}_i$; then $b - \sum c_i a_i$ belongs to $\mathcal{I}$, and its degree is $\leq \ell - 1$. By induction, we get the result.

Remark 2.3. — About the effectiveness of these constructions, we note that the only point is the effectiveness in $\mathcal{O}$; if we work with A instead of $\mathcal{D}$, $\mathcal{O}$ is replaced by $\mathbf{C}[[x]]$ and one can use Gröbner bases.

Remark 2.4. — The converse of lemma 2.2 is false: if $(a_1, \ldots, a_p)$ is a basis of $\mathcal{I}$, $(\sigma(a_1), \ldots, \sigma(q))$ is not in general a basis of $\operatorname{gr} \mathcal{I}$ (even if we take the principal symbol of the "exact degree" of the a_i's). Here is a classical counterexample: take $\mathcal{I} = (\partial_1, \delta)$, with $\delta = x_1 \partial_2 + x_2 \partial_3 + \cdots + x_{n-1} \partial_n$; then ξ_1 and $\operatorname{gr} \delta$ do not generate $\operatorname{gr} \mathcal{I}$.

Actually, one has $[\partial_1, \delta] = \partial_2$; then $\partial_2 \in \mathcal{I}$; similarly, $[\partial_2, \delta] = \partial_3$, then $\partial_3 \in \mathcal{I}$, and so on; finally, we find that $\mathcal{I}$ is the ideal generated by $\partial_1, \ldots, \partial_n$, i.e. the ideal of all differential operators without constant term; then $\operatorname{gr} \mathcal{I} = (\xi_1, \ldots, \xi_n)$.

In fact, Stafford proved that, for any n, all ideals of $\mathcal{D}$ have *two* generators ! But this is obviously false for $\operatorname{gr} \mathcal{D}$ (and it is even false if we restrict ourselves to $\operatorname{gr} \mathcal{I}$, where $\mathcal{I}$ an ideal of $\mathcal{D}$).

3.   Dimension theory

Let M be a finite (left) $\mathcal{D}$-module; the preceding constructions extend to M in the following way:

We define a *filtration* of M as a sequence $M_0 \subset \cdots \subset M_k \subset \cdots$ of finite $\mathcal{O}$-submodules of M with the following properties:

$i)$ $\cup M_k = M$

$ii)$ $\mathcal{D}_k M_\ell \subset M_{k+\ell}$ for all $k \geq 0$, $\ell \geq 0$.

The filtration is said to be *good* if one has $\mathcal{D}_k M_\ell = M_{k+\ell}$ for $\ell \geq \ell_0$ and all k ($\ell \geq \ell_0$ and $k = 1$ is sufficient).

Examples.

i) The obvious filtration of $\mathcal{D}$ (as a left $\mathcal{D}$-module) is obviously good.

ii) Any finite $\mathcal{D}$-module admits a good filtration: let $(m_1, \ldots, m_p)$ be genera-tors of M; then the map $\mathcal{D}^p \to M$ defined by $(a_1, \ldots, a_p) \mapsto a_1 m_1 + \cdots + a_p m_p$ is surjective. We just take the filtration of M as the quotient of the obvious filtration of $\mathcal{D}^p$; this is clearly a good filtration.

iii) If $\mathcal{I}$ is an ideal of $\mathcal{D}$, the filtration of $\mathcal{I}$ induced by the trivial filtration of $\mathcal{D}$ is good; it is an easy consequence of the arguments of the preceding section.

iv) More generally, one can prove, in a more or less similar way, the following result: if $N \subset M$ are finite $\mathcal{D}$-modules, any good filtration of M induces a good filtration of N ("property of the Artin-Rees type").

Now, take a finite $\mathcal{D}$-module M and a good filtration $\{M_k\}$ on it. Then $\operatorname{gr} M = \oplus(M_{k+1}/M_k)$ is a module over $\operatorname{gr}\mathcal{D} = \mathcal{O}[\xi]$, and this module is *finite* [exercise: verify that this is equivalent to the goodness of the filtration].

Let V be the *support* of $\operatorname{gr} M$, which can be defined as follows: let $\overline{\mathcal{I}}$ be the ideal of $\operatorname{gr}\mathcal{D}$ annihilating $\operatorname{gr} M$; then V is the "set" of zeroes of $\overline{\mathcal{I}}$ (more exactly, V is a germ of sets for the family of $U_x \times \mathbf{C}^n_\xi$, U an open neighbor-hood of $0 \in \mathbf{C}^n$: if we take generators $\bar{a}_1, \ldots, \bar{a}_p$ of $\overline{\mathcal{I}}$, their common zeroes define a set in $U \times \mathbf{C}^n$ for some U; and the corresponding germ is obviously independent of the chosen generators; in what follows, I will slightly abuse the language and speak of V as a set). V is analytic in all the variables (x, ξ) and algebraic and homogeneous with respect to the variables ξ. We call it the "characteristic variety" of M and denote it by $\operatorname{char} M$.

Theorem 3.1. — *The characteristic variety depends only on M, and not on the chosen filtration.*

i) First, we prove the result for two good filtrations $\{M_k\}$ and $\{M'_k\}$ which satisfy $M'_k \subset M_k \subset M'_{k+1}$. We have the exact sequences
$$0 \to M_k/M'_k \to M'_{k+1}/M'_k \to M'_{k+1}/M_k \to 0 \quad \text{and}$$
$$0 \to M'_{k+1}/M_k \to M_{k+1}/M_k \to M_{k+1}/M'_{k+1} \to 0.$$

Denote by A (resp. B) the $\operatorname{gr}\mathcal{D}$-modules $\oplus(M_k/M'_k)$ (resp. $\oplus(M'_{k+1}/M_k)$). The preceding exact sequences give exact sequences
$$0 \to A \to \operatorname{gr}' M \to B \to 0 \text{ and}$$
$$0 \to B \to \operatorname{gr} M \to A \to 0.$$

Now, well-known properties of supports give $\operatorname{supp}(\operatorname{gr}' M) = \operatorname{supp}(A) \cup \operatorname{supp}(B)$ and the same for $\operatorname{gr} M$; then the result follows.

ii) Now, let $\{M_k\}$ and $\{N_k\}$ be two good filtrations of M. There exists some ℓ such that one has $M_k \subset N_{k+\ell}$ and $N_k \subset M_{k+\ell}$ (this is an easy consequence of the goodness). The result is obvious for $\ell = 0$, since, in that case, the filtrations coincide; we will then prove the result by induction on ℓ.

For $\ell \geq 1$, put $M'_k = M_k \cap N_{k+\ell-1}$, $N'_k = N_k \cap M_{k+\ell-1}$; these new filtrations are obviously good. One has $M'_k \subset N'_{k+\ell-1}$, $N'_k \subset M'_{k+\ell-1}$; therefore, by induction, the filtrations $\{M'_k\}$ and $\{N'_k\}$ give the same characteristic variety. On the other hand, we have $M'_k \subset M_k \subset M'_{k+1}$ and the same with M replaced by N; therefore by *i)*, in the preceding assertion, we can replace $\{M'_k\}$ by $\{M_k\}$ and $\{N'_k\}$ by $\{N_k\}$; this completes the proof.

Examples.

i) If $M = \mathcal{D}/\mathcal{I}$, and we take the filtration quotient on M, then $\operatorname{gr} M = \operatorname{gr}\mathcal{D}/\operatorname{gr}\mathcal{I}$ (on $\mathcal{I}$, we take the induced filtration). V is the "set" of zeroes of $\operatorname{gr}\mathcal{I}$, or the "set" of zeroes of the principal symbols of elements of $\mathcal{I}$.

But, if $M = \mathcal{D}/\mathcal{I}'$, $\mathcal{I}'$ another ideal, we can have $\operatorname{gr}' M \neq \operatorname{gr} M$: look e.g. at the case studied in 1,ii, [...].

ii) Suppose we have an exact sequence $0 \rightarrow M' \rightarrow M \rightarrow M'' \rightarrow 0$; if we have a good filtration on M, we have seen that the induced filtration on M' is good; on the other hand, the filtration quotient on M'' is obviously good. With these choices, we have again an exact sequence $0 \rightarrow \operatorname{gr} M' \rightarrow \operatorname{gr} M \rightarrow \operatorname{gr} M'' \rightarrow 0$ (verification left as an exercise). Therefore, we have $\operatorname{char} M = \operatorname{char} M' \cup \operatorname{char} M''$.

In principle, the preceding properties could be used to determine the characteristic variety of a $\mathcal{D}$-module, e.g. by using induction on the number of generators. However, this will only work in very special cases; in general, to calculate a characteristic variety is a hard problem!

There are severe restrictions on varieties of zeroes of ideals in $\operatorname{gr}\mathcal{D}$ that are characteristic varieties of $\mathcal{D}$-modules. The general necessary and sufficient condition is not known; however, a very important condition is the *involutiveness*.

Recall that this means: if we have two functions f and $g \in \mathcal{O}[\xi]$, we define their *Poisson bracket* by the formula

$$\{f,g\} = \sum_1^n \left[\frac{\partial f}{\partial \xi_i} \frac{\partial g}{\partial x_i} - \frac{\partial f}{\partial x_i} \frac{\partial g}{\partial \xi_i} \right].$$

This is related to the theory of differential equations as follows: if a and b are elements of $\mathcal{D}$, of degree respectively k and ℓ, then $[a,b] = ab - ba$ is of degree

$\leq k + \ell - 1$ and, taking the principal symbols of the corresponding degrees, one has $\sigma[a, b] = \{\sigma(a), \sigma(b)\}$ (verification left to the reader).

Now, let $\mathcal{I}$ be an ideal of $\mathcal{O}[\xi]$, and V its variety. Let $\widetilde{\mathcal{I}}$ be the ideal of *all* elements of $\mathcal{O}[\xi]$ vanishing on V; by the so-called "Nullstellensatz", $\widetilde{\mathcal{I}}$ is the *root* of $\mathcal{I}$, i.e. the set of all $a \in \mathcal{O}[\xi]$ such that $a^k \in \mathcal{I}$ for some k.

One says that V is involutive if $\widetilde{\mathcal{I}}$ is stable under Poisson bracket, in other words if one has $\{\widetilde{\mathcal{I}}, \widetilde{\mathcal{I}}\} \subset \widetilde{\mathcal{I}}$. Then, one has the following result.

Theorem 3.2. — *If M is a $\mathcal{D}$-module, then char M is involutive.*

This theorem is quite natural, since, if $\mathcal{I}$ is an ideal of $\mathcal{D}$, one has obviously $[\mathcal{I}, \mathcal{I}] \subset \mathcal{I}$; therefore, taking the principal symbols, one has $\{\text{gr}\,\mathcal{I}, \text{gr}\,\mathcal{I}\} \subset \text{gr}\,\mathcal{I}$; but this does not prove the result at all: what we want is $\{\widetilde{\text{gr}\,\mathcal{I}}, \widetilde{\text{gr}\,\mathcal{I}}\} \subset \widetilde{\text{gr}\,\mathcal{I}}$, $\widetilde{\text{gr}\,\mathcal{I}}$ the root of gr $\mathcal{I}$, and this is more difficult.

All the proofs use microlocalization, in one way or another; I refer to the literature.

An important consequence of this theorem is the following: if an analytic set X is involutive, at each of its points its dimension is $\geq n$ (see appendix at the end of this section; I recall that the smooth part X^{reg} of X, i.e. the set of points where X is a non-singular complex manifold, is dense in X; and the dimension of X is by definition the complex dimension of X^{reg}). If we call the dimension of char M the *dimension of M*, then we have

Corollary 3.3. — *If M is a $\mathcal{D}$-module, its dimension is $\geq n$.*

A very interesting case is the case where dim $M = n$; in that case, M is said to be *holonomic* and its characteristic variety *lagrangian* (see appendix for this last notion).

Examples of holonomic modules.

i) $\mathcal{O}$ as $\mathcal{D}$-module: in fact, we have a surjection $\mathcal{D} \to \mathcal{O}$ by $a \mapsto a(1)$. The kernel is the ideal generated by $(\partial_1, \dots, \partial_n)$; therefore, char $\mathcal{O}$ is the set $\xi_1 = \cdots = \xi_n = 0$.

ii) $\mathcal{D}/\mathcal{I}$, $\mathcal{I}$ an ideal generated by $x_1, \dots, x_n$; the canonical generator (= image of $1 \in \mathcal{D}$) is usually called the "Dirac function" and denoted by δ, because the Dirac "function", or better distribution, is precisely annihilated by the x_i's. Of course, here, we have an "abstract" Dirac function, which is not in any sense a function or a distribution.

iii) For $f \in \mathcal{O}$, $f \neq 0$, one can prove that $\mathcal{O}[f^{-1}]$ is holonomic (see section 1, ii)). More generally, if M is holonomic, $M[f^{-1}]$ is holonomic. This is a difficult result due to Kashiwara. But the determination of the characteristic variety of $M[f^{-1}]$ is a very hard problem, only solved at the moment (by Ginzburg and Sabbah independently) in the case where M has "regular singularities". Note also that, if M is only finite over $\mathcal{D}$, then $M[f^{-1}]$ is not finite in general; counterexample, $M = \mathcal{D}$.

Appendix to section 3 : Symplectic geometry and involutive varieties

Let U be an open set in $\mathbf{C}^n$, with coordinates $(x_1, \ldots, x_n)$; an analytic subset W of $U \times \mathbf{C}^n$ is called *involutive* if the following condition is satisfied: let $a \in V$, and let f and g be two holomorphic functions near a, vanishing on V; then $\{f, g\}$, defined as before, vanishes also on W.

The definition given here seems a little bit more restrictive than the one used before, since here we work with *local* equations and not only with global ones; but, if M is a finite $\mathcal{D}$-module, then char M is also involutive in the stronger sense: this would follow from the general theory of holomorphic functions; but, actually the proof of 3.2 gives the stronger result directly.

Let $a \in U \times \mathbf{C}^n$; for $\alpha, \alpha' \in T_a^*(U \times \mathbf{C}^n)$, the space of differentials at a, we define $\{\alpha, \alpha'\}_a$ as follows.

If $a = \sum_1^n \alpha_i dx_i + \sum \beta_i d\xi_i$, $\alpha' = \sum(\alpha_i' dx_i + \beta_i' d\xi_i)$, we put $\{\alpha, \alpha'\}_a = \sum(\beta_i \alpha_i' - \alpha_i \beta_i')$; therefore, for f and g holomorphic near a, we have $\{f, g\}(a) = \{df, dg\}_a$.

Now, take an involutive analytic set $W \subset U \times \mathbf{C}^n$; if a is a point of W^{reg}, the regular part of W, the conormal N_a at a to W (i.e. the subset of $T_a^*(U \times \mathbf{C}^n)$ for α satisfying $\alpha|W = 0$) is generated by differential of functions vanishing on W, because of the definition of W^{reg}; therefore, for $\alpha, \alpha' \in N_a$, one has $\{\alpha, \alpha'\}_a = 0$. Using the notation $N_a^\perp$ orthogonal of N_a for the bilinear alternating form $\{\ ,\ \}_a$, this means that one has $N_a^\perp \subset N_a$. Now, as this form is obviously nondegenerate, one has $\dim N_a^\perp = 2n - \dim N_a$. Therefore one has $\dim N_a \leq n$, on $\dim_a W \geq n$. This proves the statement on the dimension used at corollary 3.3.

There is another way to look at this question, which is a little bit longer, but more illuminating. Put $\lambda = \sum \xi_i dx_i$, $\omega = d\lambda = \sum d\xi_i \wedge dx_i$.

For $a \in U \times \mathbf{C}^n$, ω gives an alternating two-form Ω on $T_a(U \times \mathbf{C}^n)$, by $\Omega(X, Y) = \langle X \wedge Y, \omega \rangle$; this form is non-degenerate, and therefore gives an isomorphism $\tilde{\Omega} : T_a(U \times \mathbf{C}^n) \to T_a^*(U \times \mathbf{C}^n)$ by $\langle \tilde{\Omega}(X), Y \rangle = \Omega(X, Y)$ or equivalently $\tilde{\Omega}(X) = X \lrcorner \omega$ ("$\lrcorner$", is the interior product).

If f is a holomorphic function on $U \times \mathbf{C}^n$, the vector field $H_f = \tilde{\Omega}^{-1}(df)$ is called the Hamiltonian vector field of f: this is the field corresponding

to the Hamilton equation with the "energy" f, which is explicitly given by
$H_f = \sum(\frac{\partial f}{\partial x_i}\frac{\partial}{\partial \xi_i} - \frac{\partial f}{\partial \xi_i}\frac{\partial}{\partial x_i})$.

Then, one has $\{f, g\} = -\langle H_f, dg \rangle$; in other words, the bilinear form $\{\ ,\ \}_a$ introduced above is, up to the sign, the image of Ω by the isomorphism $\tilde{\Omega}$. Now, the involutiveness can be interpreted naturally in term of tangent space; if W in an involutive analytic subset of $U \times \mathbf{C}^n$, take $a \in W^{\mathrm{reg}}$; as $T_a W$ is the orthogonal to N_a (for the duality vectors $\leftrightarrow$ forms), the relation $N_a^\perp \supset N_a$ is translated here into $T_a W^\perp \subset T_a W$, the "$\perp$" referring to the form Ω; therefore, one has $\dim T_a W \geq n$; if we have equality, which means that W is lagrangian at a, we have $T_0 W^\perp = T_a W$; this means that Ω vanishes when restricted to $T_a W$; therefore, if W is lagrangian, $\omega | W^{\mathrm{reg}} = 0$. Furthermore, if W is homogeneous w.r. to the ξ variable, which is the case for the characteristic varieties of $\mathcal{D}$-modules, one shows that one has also $\lambda | W^{\mathrm{reg}} = 0$.

The preceding construction has also the interest of being compatible with the natural way to change the coordinates. Actually, it is easy to prove that the principal symbols of differential operators are transformed as functions on the cotangent space T^*U; in coordinates $(x_1, \dots, x_n)$, one has an identification of T^*U with $U \times \mathbf{C}^n$ by identifying $\sum \alpha_i dx_i \in T_a^*U$ with the point (a, α), $\alpha = (\alpha_1, \dots, \alpha_n)$; but this means just that the "Liouville form" λ is canonically defined on T^*U; therefore the other construction ω, $\{\ ,\ \}$, etc... . are also canonical, i.e. independent of the choice of coordinates.

The simplest examples of Lagrangian manifolds are the *conormal* manifolds: if X is a submanifold of U, then $N_X = \{(a, \alpha) \mid a \in X, \alpha \in T_a^*U, \alpha | X = 0\}$ is lagrangian. After a change of coordinates, we can suppose, at least locally, that $X = \{x_1 = \dots = x_p = 0\}$; then $N_X = \{x_1 = \dots = x_p = 0; \xi_{p+1} = \dots = \xi_n = 0\}$. One can prove that all *homogeneous* lagrangian varieties are adherences of locally finite unions of such conormal manifolds.

4.   Holonomic modules in one variable

Here $n = 1$; the variable is denoted by x. Let M be a holonomic $\mathcal{D}$-module. Suppose first that M is cyclic, i.e. of the form $\mathcal{D}/\mathcal{I}$ [actually, one can prove that this is always the case, but we will not use this result here]. An operator $a \in \mathcal{I}$ has the form $a = \sum_0^n a_k \partial^k$, $a_k \in \mathcal{O} = \mathbf{C}\{x\}$; suppose $a_n \neq 0$. We put $d(a) = m$, $v(a) = v(a_n)$, the *valuation*, or order of zero at 0, of a_n. The principal symbol of a is $a_n \xi^n$, or, up to an invertible factor $x^m \xi^n$, $m = v(a)$. Its set of zeroes is either $X = \{\xi = 0\}$, or $T_0^*X = \{x = 0\}$, or $X \cup T_0^*X$. Therefore, for the characteristic variety of M, we have only four possibilities (if we exclude the trivial case $M = \{0\}$).

i) char $M = X \cup T_0^*X$

ii) char $M = T_0^* X$

iii) char $M = X$

iv) char $M = \{0\}$.

We note that the first three cases are involutive, and case *iv)* not: therefore this case is excluded by theorem 3.2; I shall re-prove this result later (see also the paper by P. Maisonobe).

Ignoring the fact that all holonomic $\mathcal{D}$-modules are cyclic, we can prove easily the same results in general, by induction on the number of generators.

Here, I shall study cases *ii)* and *iii)*, and I shall say only a few words on the "general" case *i)*.

Case ii): char $M = T_0^* X$

This means that for a good filtration of M, the support of gr M is $\{x = 0\}$; by the "Nullstellensatz" for any $\bar{m} \in$ gr M, one has $x^k \bar{m} = 0$ for some k; but this implies immediately that the same is true for M.

We first consider the special case where $M = \mathcal{D}/\mathcal{D}x$; let δ denote the canonical generator of M (image of $1 \in \mathcal{D}$). Then it is immediately verified that every element of M can be written in a unique way $m = \sum \alpha_k \partial^k \delta$; $\alpha_k \in \mathcal{O}$: indeed, for $a \in \mathcal{D}$, write $a = \sum \partial^k a_k$, $a_k \in \mathcal{O}$; then $a_k = \alpha_k + x b_k$, with $\alpha_k = a_k(0)$, $b_k \in \mathcal{O}$, and the result follows.

Now, take any holonomic M with support $T_0^* X$. Put $E = \{m \in M \mid xm = 0\}$. We have an obvious map of left $\mathcal{D}$-module $\mathcal{D}\delta \otimes_{\mathbb{C}} E \to M$.

Proposition 4.1. — *This map is bijective.*

a) Injectivity. — Suppose that we have a relation $\sum a_i e_i = 0$, $a_i \in \mathcal{D}$, $e_i \in E$; then, we have $\sum x a_i e_i = 0$ and $\sum a_i x e_i = 0$, then $\sum [a_i, x] e_i = 0$. But, if $a_i = \sum a_{i_k} \partial^k$, we have $[a_i, x] = \sum k a_{i,k} \partial^{k-1}$ (Leibniz rule). Therefore, we have a new relation whose coefficients have degree reduced exactly by one; by induction, we find that the e_i's are not linearly independent on $\mathbb{C}$.

b) Surjectivity. — Take $m \in M$. There is a k such that $x^k m = 0$; suppose $k \geq 2$; we prove by induction on k that m belongs to $\mathcal{D}E$, the submodule generated by E. Put $xm = n_1$, $x \partial m + km = n_2$ then, one has $x^{k-1} n_1 = 0$ (obvious), and also $x^{k-1} n_2 = \partial(x^k m) = 0$. Therefore, by induction, we can suppose that n_1 and n_2 belong to $\mathcal{D}E$. But we have $n_2 - \partial n_1 = x \partial m + km - \partial(xm) = (k-1)m$; this implies the result.

Note that this result implies the following: the $\mathcal{D}$-sub-modules of M are in one-to-one correspondence with the submodules over $\mathbf{C}$ of E; in particular, E is finite over $\mathbf{C}$ (otherwise, there would be an infinite sequence of strictly increasing modules of M, which would not be nœtherian). This proves also that $\mathcal{D}\delta$ is *simple*, i.e. has no non-trivial submodules. Finally, this re-proves also the fact that no finite $\mathcal{D}$-module can have $\{0\}$ as characteristic variety.

Case iii): char $M = X$

The prototype here is $\mathcal{O} = \mathcal{D}/\mathcal{D}\partial$ (the bijection is given, as we have seen in section 1, by $a \mapsto a(1)$, $a \in \mathcal{D}$, or equivalently by $a = \sum a_k \partial^k \mapsto a_0$).

In the general case, the result is similar to the preceding one, but the proof is rather different. Let M be a $\mathcal{D}$-module with char $M = X$, and put $E = \{m \in M \mid \partial m = 0\}$.

Proposition 4.2. — *The natural map of $\mathcal{D}$-modules $\mathcal{O} \otimes_{\mathbf{C}} E \to M$ is bijective.*

We will prove directly that M is isomorphic, as $\mathcal{D}$-module, to $\mathcal{O}^p$ for some p (this easily implies the proposition).

a) M is finite over $\mathcal{O}$; indeed, take a good filtration of M; take generators $\bar{m}_i$ of gr M, say homogeneous of degree $\le \ell$; there exists k such that $\xi^k \bar{m}_i = 0$. Therefore, one has $\mathrm{gr}^n M = 0$ for $n \ge k + \ell$; but this implies that $M = M_{k+\ell}$, and the result follows.

b) M is free over $\mathcal{O}$ ($= M$ is isomorphic to $\mathcal{O}^p$ as $\mathcal{O}$-module). Let $\mathfrak{m}$ be the maximal ideal of $\mathcal{O}$, generated by x, and let $\bar{m}_1, \dots, \bar{m}_p$ be a basis of $M/\mathfrak{m}M$ over $\mathbf{C}$; lift $\bar{m}_1, \dots, \bar{m}_p$ to $m_1, \dots, m_p \in M$. The result follows from the next two lemmas.

Lemma 4.3. — *The m_i's generate M over $\mathcal{O}$.*

This is a special case of the so-called "Nakayama lemma". Let M' denote the submodule generated by the m_i, and put $N = M/M'$; then N is finite over $\mathcal{O}$ and satisfies $N = \mathfrak{m}N$; this implies $N = 0$ [take n_i, generators of N; one has $n_i = \sum a_{ij} n_j$, $a_{ij} \in \mathfrak{m}$; then the matrix $\mathrm{id} - (a_{ij})$ is invertible, and so $n_i = 0$].

Lemma 4.4. — *The m_i's are linearly independent over $\mathcal{O}$.*

Let $\sum a_i m_i = 0$ be a relation, and let $d = \inf v(a_i)$. By hypothesis, one has necessarily $d \ge 1$, otherwise the $\bar{m}_i$ would not be independent over $\mathbf{C}$.

Differentiating the relation we find $\sum(\partial a_i)m_i + \sum a_i \partial m_i = 0$; expressing the ∂m_i's in terms of the m_i's, we find a relation of "degree" $d - 1$. By induction, this gives the result. [Actually, we could have used a different argument, instead of these lemmas: it follows from what we have seen in ii) that M has no torsion element, i.e. no element annihilated by x^k; and, as well is known, this implies that M is free over $\mathcal{O}$. But this argument is not so easily extended to several variables as the previous one.]

c) Finally, to prove the result, we write the equations $\partial m_j = \sum a_{ij} m_i$, $a_{ij} \in \mathcal{O}$; in short $\partial \underline{m} = \underline{m} A$, $\underline{m} = (m_1, \dots, m_p)$, $A = (a_{ij})$. The change of basis $\underline{n} = \underline{m} S$, $S \in G\ell(p, \mathcal{O})$, gives similar equations for $\underline{n}$, with A replaced by $S^{-1} A S + S^{-1} \frac{dS}{dx}$. By the existence theorem for differential equations we can take S satisfying $\frac{dS}{dx} + AS = 0$, $S(0) = \text{id}$. Then, one has $\partial n_i = 0$ and the result is proved.

Note that, again, these arguments re-prove the fact that $\{0\}$ cannot be a characteristic variety.

I will end this section by a few words on the "general" case (case i)); see also some other information, related standard basis, in the paper by P. Maisonobe. Denote by $K = \mathcal{O}[x^{-1}]$ the *field* of meromorphic functions at $0 \in \mathbf{C}$; we will use the name "meromorphic connection" for a finite dimensional vector space L over K with an action of ∂ satisfying the following properties:

$$\begin{cases} i) & \partial \text{ is } \mathbf{C}\text{-linear;} \\ ii) & \text{for } \varphi \in K, \ell \in L, \text{ one has } \partial(\varphi \ell) = \frac{d\varphi}{dx} \ell + \varphi \partial \ell. \end{cases}$$

These properties are equivalent to the fact that $\mathcal{D}[x^{-1}]$, the ring of differential operators with meromorphic coefficients, acts on L; in particular, L is a left $\mathcal{D}$-module. Then, one has the following result.

Proposition 4.5. — *The meromorphic connections are the holonomic $\mathcal{D}$-modules M on which x acts bijectively.*

i) Suppose that M is a holonomic $\mathcal{D}$-module on which x is bijective. Then $\mathcal{O}[x^{-1}] = K$ acts on M, and even $\mathcal{D}[x^{-1}]$; and one easily verifies i) and ii). Therefore, it is sufficient to prove that M is finite over K.

Take $m \in M$; we will show that $\mathcal{D}m[x^{-1}]$ is finite over K (ignoring the fact that all holonomic modules are cyclic, we will deduce the result by induction on the number of generators of M). There is an $a \in \mathcal{D}$ such that $am = 0$, otherwise $\mathcal{D}m$ would not be holonomic; by multiplying a by a suitable power of x, we can replace it by $a' = \partial^n + \sum a_i \partial^{n-i}$, $a_i \in K$. Therefore $\mathcal{D}m[x^{-1}]$ is a quotient of $\mathcal{D}[x^{-1}]/\mathcal{D}[x^{-1}]a'$; but it is immediate that this last space has dimension n over K; this proves the result.

ii) Let L be a meromorphic connection; L is obviously a finite $\mathcal{D}[x^{-1}]$-module, and we want to prove that it is finite also over $\mathcal{D}$. Again, we can suppose L cyclic and isomorphic to $\mathcal{D}[x^{-1}]/\mathcal{D}[x^{-1}]a$. Here, by multiplying a by a suitable power of x, we can suppose that one has $a = b(x\partial) + xq(x, x\partial)$, with $b \in \mathbf{C}[T]$, $b \neq 0$ (b actually corresponds to the "horizontal side" of the Newton polygon of a). If we replace the generator ℓ satisfying $a\ell = 0$ by $x^k\ell$, we replace a by $b(x\partial - k) + xq(x, x\partial - k)$; therefore, taking $k \ll 0$, we can suppose that $b(m) \neq 0$, $m \in \mathbf{N}$. Now, with this hypothesis, we will show that x acts bijectively on $\mathcal{D}/\mathcal{D}a$; or, in other words, that $\mathcal{D}/\mathcal{D}a$ is isomorphic with its localization $\mathcal{D}[x^{-1}]/\mathcal{D}[x^{-1}]a$. This will prove the result.

To prove that x acts bijectively on $\mathcal{D}/\mathcal{D}a$ is equivalent to proving that the right multiplication $\cdot a$ acts bijectively on $N = \mathcal{D}/x\mathcal{D}$ [exercise]. In order to prove this last result, we write N in a similar way as we have done for $\mathcal{D}/\mathcal{D}x$: denoting the generator by $\check{\delta}$, all elements are written in a unique way $\sum c_k \check{\delta}\partial^k$; if we filtrate N by "k", we find that the action of $\cdot a$ on gr N is given by $\cdot b(x\partial)$; but, one has $\check{\delta}\partial^k x\partial = \check{\delta}[x\partial^{k+1} + k\partial^k] = k\check{\delta}\partial^k$; therefore $\check{\delta}\partial^k b(x\partial) = b(k)\check{\delta}\partial^k$; as $b(k) \neq 0$, $k \in \mathbf{N}$, the action of $b(x\partial)$ is bijective on gr N and the result follows. This proves the proposition.

The preceding result easily implies the following one, which is the special case for dimension one of Kashiwara's result mentioned above: if M is holonomic, $M[x^{-1}]$ is holonomic. Now, the kernel and the cokernel of the map $M \to M[x^{-1}]$ have support at 0; therefore, according to 4.1, they are isomorphic to $(\mathcal{D}\delta)^p$ for some p. On the other hand, $M[x^{-1}]$ is a meromorphic connection, and the structure of meromorphic connection is in principle well-known by the theory of formal decomposition + Stokes phenomenon; I will not recall this here. Finally, we see that in one variable, the local theory of holonomic $\mathcal{D}$-modules reduces to the theory of meromorphic connections up to extensions by "trivial" factors $\mathcal{D}\delta$. For several variables, the situation is more difficult.

Remark 4.6. — The proof of Kashiwara's theorem in the general case uses b-functions and their functional equation (see the paper by P. Maisonobe). The proof given here in the case of dimension one is actually a special case of this argument; in fact, one can prove that an equation of the form $b(x\partial)\ell + xq(x, x\partial)\ell = 0$ is equivalent, up to x-torsion, to an equation $b[-(s+1)]x^s\ell + r(s, x, \partial)x^{s+1}\ell = 0$ (the "b" is the same in both equations). In particular, looking successively the cases $s = -1, -2, \ldots$, we find expressions for the $x^{-k}\ell$ in terms of ℓ and its derivatives, when b has no non-negative integral root.

Here is a short way to go from the first formula to the second (but this is not very illuminating: the most natural way would use "direct images" and $\delta(t-x)$, but it would be longer to explain).

Put $a = b(x\partial) + xq(x, x\partial)$; let ℓ be the canonical generator of $L = \mathcal{D}/\mathcal{D}a$, and denote by $M \subset L[s, x^{-1}]x^s$ the set of elements of the form $p(s, x, \partial)(\ell x^{s+1})$,

$p \in \mathcal{D}[s]$. We have $(x\partial\ell)x^s + (s+1)\ell x^s = \partial(\ell x^{s+1})$, and therefore

$$[(x\partial)^k\ell]x^s + (s+1)[(x\partial)^{k-1}\ell]x^s = \partial[((x\partial)^{k-1}\ell)x^{s+1}].$$

Applying this formula first with s replaced by $s+1$, we find by induction that one has $[(x\partial)^k\ell]x^{s+1} \in M$, and therefore $[xq(x, x\partial)\ell]x^s \in M$.

Applying the same formula with s, we find $[(x\partial)^k\ell]x^s + (s+1)[(x\partial)^{k-1}\ell]x^s \in M$, and therefore, by induction, $[b(x\partial)\ell]x^s - b[-(s+1)]\ell x^s \in M$. Then, from $b(x\partial)\ell + xp(x, x\partial)\ell = 0$, we deduce $b[-(s+1)]\ell x^s \in M$, which is the required result.

5.   Operations on $\mathcal{D}$-modules

I shall speak here of the case of $A = \mathbf{C}[x, \partial]$, since it is simpler to explain (the analytical case requires us to work with sheaves, at least for direct images). As I said before, the dimension theory applies here: we filtrate A by the subspaces A^k of operators of degree $\leq k$ in ∂; then, for a finite (left) A-module M, one defines good filtration, characteristic variety, and dimension as we did for $\mathcal{D}$; assertions (3.1), (3.2) and (3.3) are still valid.

Then, the basic operations are the following ones.

i) Localization

Let $f \in \mathbf{C}[x]$, $f \neq 0$. Then Kashiwara's theorem is also true here: if M is holonomic on A, $M[f^{-1}]$ is also holonomic (but, if M is finite, $M[f^{-1}]$ need not be finite).

ii) Inverse image

There are two basic cases:

α) The projection $p = \mathbf{C}^{n+1} \to \mathbf{C}^n$ defined by $p(x_1, \ldots, x_n, t) = (x_1, \ldots, x_n)$. The idea here is to mimic the inverse image of functions: if M is an A-module (in dimension n) with generator e, we write p^*M for the module generated by an element, denoted by p^*e, which satisfies the same equations plus the equation $\partial_t(p^*e) = 0$. This is formalized in the following general definition: one takes $p^*M = M \otimes_{\mathbf{C}} \mathbf{C}[t]$, with the following action: $a(x, \partial_x)(m \otimes n) = a(x, \partial_x)m \otimes n$, $b(t, \partial_t)(m \otimes n) = m \otimes b(t, \partial_t)n$ (the preceding example is obtained by the identification $p^*e = e \otimes 1$).

In this case, everything goes very simply; if M is finite, then p^*M is finite and $\operatorname{char} p^*M = \{(x, t, \xi, 0) \mid (x, \xi) \in \operatorname{char} M\}$. Therefore, one has $\dim p^*M = \dim M + 1$; in particular, if M is holonomic, then p^*M is holonomic.

These results are easy (hint: if M_k is a good filtration of M, filtrate $p^* M$ by $M_k \otimes_{\mathbf{C}} \mathbf{C}[t]$).

β) *The embedding* $i : \mathbf{C}^n \to \mathbf{C}^{n+1}$ defined by $i(x_1, \ldots, x_n) = (x_1, \ldots, x_n, 0)$. Similar considerations lead to the following definition: we take $i^* M = M/tM$ (t, the last coordinate in $\mathbf{C}^{n+1}$), and we make $a(x, \partial_x)$ act on it in the obvious way.

γ) *The general case* is reduced to the preceding cases by the usual procedure: let f be a polynomial map $\mathbf{C}^m \to \mathbf{C}^n$; we factorize f by $\mathbf{C}^m \xrightarrow{g} \mathbf{C}^{m+n} \xrightarrow{h} \mathbf{C}^n$, $g = (\mathrm{id}, f)$, h the projection. We define h^* by induction on m, using α). To define g^* we make the change of coordinates $z_i = y_i - f_i(x)$, then apply β) by induction. One has the following result.

Theorem 5.1. — *If M is holonomic, then $f^* M$ is holonomic.*

But if M is finite $f^* M$ is not necessarily finite. Furthermore, there is in general no simple relation between the characteristic variety of M and that of $f^* M$. The preceding theorem reduces by induction to case β). this was first proved by Bernstein.

Remark 5.2. — Actually this definition is not general enough; one would have to consider also "higher restrictions"; for instance, in the case β), one should also consider $\ker(t, M)$ and not only $\operatorname{coker}(t, M)$.

iii) Direct image

Here again, there are two basic cases; Using the same notations as in *ii)*.

α) *The projection* $p : \mathbf{C}^{n+1} \to \mathbf{C}^n$. The idea here is to mimic "integration in the fiber"; so we take $p_* M = M/\partial_t M$, with the obvious action of $\mathbf{C}[x, \partial_x]$.

Here the result (Bernstein) is

Theorem 5.3. — *If M is holonomic, $p_* M$ is holonomic.*

Again, if M is finite, $p_* M$ need not be finite.

β) *The embedding* $i : \mathbf{C}^n \to \mathbf{C}^{n+1}$. Denote by Δ the $\mathbf{C}[t, \partial_t]$-module

$$\mathbf{C}[t, \partial_t]\delta = \mathbf{C}[t, \partial_t]/\mathbf{C}[t, \partial_t]t .$$

One takes here $i_* M = M \otimes_{\mathbf{C}} \Delta$, with the obvious action of a: the x and ∂_x act on the first factors, the t and ∂_t on the second.

Here one easily verifies the following result, left as an exercise: if M is finite, i_*M is finite; and char $M = \{(x, 0, \xi, \tau) \mid (x, \xi) \in \text{char } M\}$. In particular, if M is holonomic, i_*M is holonomic.

γ) The general case reduces as in ii) to cases α) and β). In particular, theorem 5.3 is true for general polynomial mappings.

Remark 5.4. — In the situation of case β), one has the following result due to Kashiwara: *the functor $M \mapsto i_*M$ is an equivalence* "$\mathbf{C}[x, \partial_x]$-finite modules" $\longleftrightarrow$ "$\mathbf{C}[x, t, \partial_x, \partial_t]$-finite modules with support on $t = 0$" [i.e. such that all elements are annihilated by some power of t). A quasi-inverse is given by $N \mapsto \ker(t, N)$, the higher inverse image.

The proof is an extension of the proof that we have given in section 4, when studying the modules with support 0.

Note also the following: if λ is the canonical map $M \to M[t^{-1}]$, there is a canonical isomorphism $\ker(t, M[t^{-1}]/\lambda M) \simeq M/tM$. Therefore, theorem 5.1 is actually a consequence of the preceding result and of the theorem of Kashiwara recalled in i).

iv) Fourier transform

This operation is specific to the case of A, and does not extend to the other cases. A is just the ring of non-commutative operators in the $2n$ variables x_i, ∂_i, with the commutation relations $[x_i, x_j] = 0$, $[\partial_i, \partial_j] = 0$, $[\partial_i, x_j] = \delta_{ij}$. Therefore it can be considered as a ring of differential operators in another way, by writing $y_i = \partial_i$, $x_i = -\partial/\partial y_i$. Let M be an A-module: the same module, but interpreted in this way, we call the *Fourier transform* of M denoted by $\mathcal{F}M$.

The relations between M and $\mathcal{F}M$ when interpreted as systems of differential equations, are far from being obvious; for instance, there is no simple relation between char M and char $\mathcal{F}M$; also, the relations between the *solutions* of the systems they define, in the sense of section 1, are well understood only in dimension one. However, one has the following result.

Theorem 5.5 (Bernstein). — *One has* $\dim \mathcal{F}M = \dim M$; *in particular, $\mathcal{F}M$ is holonomic iff M is holonomic.*

The proof uses the *Bernstein* filtration of A, which is the filtration by the total degree with respect to the x_i and ∂_i; if one has a finite A-module, provided with a good filtration in this sense, one can define the Hilbert-Samuel function and the corresponding dimension (see the lectures by A. Galligo). Then, one proves that this dimension, obviously invariant under Fourier transform, is equal to the dimension of M in the former sense.

More generally, we could consider transformations $y = Ax + B\partial$, $\delta = Cx + D\partial$; the condition for the y's and the δ's to satisfy the Heisenberg commutation relation is that the matrix $\begin{pmatrix} A & B \\ C & D \end{pmatrix}$ is *symplectic*. In this case, the preceding theorem is also true, with the same proof.

6.   References

i) Here are some basic books; references to the original papers can be found there.

J.E. BJÖRK, *Rings of differential operators*, North-Holland, 1979.

A. BOREL et al., *Algebraic $\mathcal{D}$-modules*, Persp. Math. 2, Acad. Press, 1987.

Z. MEBKHOUT, *Le formalisme des six opérations de Grothendieck pour les $\mathcal{D}_X$-modules holonomes*, Hermann, 1989.

ii) For holonomic modules in one variable see the following book, and the references in it:

B. MALGRANGE, *Équations différentielles à coefficients polynomiaux*, Progress in Math. 96, Birkhäuser, 1991.

$\mathcal{D}$-modules: an overview towards effectivity

Philippe Maisonobe *

Contents

1. Differential systems in one variable   **23**

    1.1. Introduction 23

    1.2. Standard basis of a left ideal 23

    1.3. Solutions of a differential system 27

       1.3.1. $Sol(I; \mathcal{F})$ 28

       1.3.2. Regularity and irregularity 31

       1.3.3. The pair E $\rightleftarrows$ F associated to an ideal 34

    1.4. Holonomic $\mathcal{D}$ modules in one variable 35

2. Holonomic $\mathcal{D}$ modules   **40**

    2.1. Introduction 40

    2.2. Structure of coherent $\mathcal{D}$ modules supported by a point 40

    2.3. A $\mathcal{D}$ module whose characteristic variety is reduced to $T_Y^* \mathbb{C}^n$ (Y smooth) 43

    2.4. Introduction to Bernstein polynomials 46

    2.5. Computation of b_f when f has an isolated singularity 47

    2.6. Basics on regularity in several variables 52

*Laboratoire C.N.R.S. "J.A. Dieudonné", Université de Nice, Parc Valrose - B.P. 71, F-06108 NICE CEDEX 2. Partially supported by Posso, EEC ESPRIT, BRA contrat 6846.

Abstract

This paper presents the $\mathcal{D}$ module approach for the "geometric" study of the solutions and "generalized" solutions of a system of differential operators with holomorphic coefficients. One associates to such a system the quotient ring $\mathcal{D}/I$ of the ring $\mathcal{D}$ by a left ideal I and studies the properties of the complexes $\mathbf{R}hom(\mathcal{D}/I, \mathcal{F})$ (see subsection 1.4).

If we have a finite representation of the coefficients, for instance when they are polynomials, many results developed below are effective. Also many constructions which look rather abstract can be mechanized, in particular through standard basis computations (see [B.M.] [C] [G]). However the complexity of these algorithms is too high and makes them intractable in practice. The result of [Gr] shows that the membership problem has a double exponential complexity. A more geometric theory with simple exponential complexity (like in commutative algebra) does not exist yet and should be developed ...

The first part of the paper presents in detail several constructions in the one variable case. The second part provides an introduction to holonomic $\mathcal{D}$ modules and Bernstein polynomials, it relies on the paper of B. Malgrange "Motivations and introduction to the theory of $\mathcal{D}$ module" which is published in this volume. We give an algorithm and a new bound for the computation of Bernstein's polynomial in the case of an isolated singularity.

Je remercie A. Galligo et T. Iarrobino qui m'ont aidé dans la rédaction de ce manuscrit.

1. Differential systems in one variable

1.1. Introduction

Let $P = \sum c_\alpha \partial^\alpha$ be a linear differential operator with holomorphic coefficients. One may wish to derive from the formula of P the analytic properties of P, for example: the dimension of the solution spaces of P, the form of the solutions... Theorems by Cauchy and by Malgrange [M 1] give the dimension of the solution space. If P is a regular operator [W] in the sense of Fuch, the question has been extensively researched and we have some information. In order to generalize these results, one associates to P a Newton polygon and expresses the solutions of P relative to this polygon [M 4][S][R].

One can obtain analogous results for a system of linear differential equations:

$$P_1 u = 0, \quad \ldots, \quad P_k u = 0 \ .$$

To solve this problem, Briançon and Maisonobe defined the notion of standard basis of a left ideal of a differential operator [B.M.].

In the next subsection, we recall the definition and principal properties of standard bases of left ideals of differential operators. In the third subsection, we give some results on the solutions of systems of differential operators.

In the fourth subsection, we define left holonomic $\mathcal{D}$-modules and describe their basic properties. $\mathcal{D}$ denotes the ring of differentials operators.

In the following references [Bjo][Bou] [G.M.] [K 1][M 4][Mai] [P][S] ... the reader can find additional information.

1.2. Standard basis of a left ideal

We begin with some notation.

$\mathbf{C}\{x\}$ will denote the germ at the origin of holomorphic functions over $\mathbf{C}$. Let

$$g(t) = \sum_{i \in \mathbf{N}} g_i x^i$$

be a non-zero element of $\mathbf{C}\{x\}$. The valuation $val(g)$ of g is the following integer:

$$val(g) = min\{i \in \mathbf{N} \quad ; \quad g_i \neq 0\} \ .$$

$\mathcal{D}$ will denote the ring of linear differential operators in one variable with coefficients in $\mathbf{C}\{x\}$. ∂_x will be the usual derivative with respect to x.

$\mathcal{D}$ is a non-commutative ring. The commutator of ∂_x and x is the multiplication by a constant function:

$$[\partial_x, x] = \partial_x x - x \partial_x = 1 \ .$$

P a non-zero element of $\mathcal{D}$ can be written uniquely

$$P = \sum_{i=0}^{k} a_i(x)\partial_x^i \quad , \quad a_i(x) \in \mathbf{C}\{x\} \quad , a_k \neq 0 \quad .$$

Definition 1.1 *The degree of P is the integer* $\deg(P) = k$.
The valuation of P is the integer $\mathrm{val}(P) = \mathrm{val}(a_k(x))$.
The exponent of P is the pair of integers $\exp(P) = (\mathrm{val}(P), \deg(P))$.

Consider two differential operators

$$P = \sum_{i=0}^{k} a_i(x)\partial_x^i \quad \text{and} \quad Q = \sum_{i=0}^{l} b_i(x)\partial_x^i \quad ;$$

then the product PQ is

$$PQ = a_k(x)b_l(x)\partial_x^{k+l} + \sum_{i=0}^{l+k-1} c_i(x)\partial_x^i \quad .$$

We have

$$\exp(PQ) = \exp(P) + \exp(Q) \quad .$$

Proposition 1.1 *(Eucidean division by an operator)*
Let $P \in \mathcal{D} - \{0\}$, let $S \in \mathcal{D}$. There exists a unique pair (Q, R) of elements of $\mathcal{D}$ such that
$$S = QP + R$$
with
$$R = \sum_{0 \leq k < \mathrm{val}(P), \deg(P) < l \leq \deg(S)} r_{k,l} x^k \partial_x^l + R'$$
with
$$r_{k,l} \in \mathbf{C} \quad , \quad R' \in \mathcal{D} \quad , \quad \deg(R') < \deg(P) \quad .$$

Proof :   By induction. If $\deg(S) < \deg(P)$, we take $Q = 0$ and $R = R' = S$.
If $\deg(S) \geq \deg(P)$, we let $l = \deg(S)$, $m = \deg(P)$,

$$S = s_l(x)\partial_x^l + \ldots + s_0(x) \quad , \quad P = a_m(x)\partial_x^m + \ldots \quad ,$$

we divide $s_l(x)$ by $a_m(x)$ (as series), and we get

$$s_l(x) = q(x)a_m(x) + \sum_{0 \leq j \leq \mathrm{val}(P)-1} r_j x^j \quad \text{with} \quad r_j \in \mathbf{C} \quad .$$

Now, we consider

$$S_1 = S - (\sum_{j=0}^{\text{val}(P)-1} r_j x^j)\partial_x^l - q(x)\partial_x^{l-p}P \quad ,$$

which satisfies $\deg(S_1) < \deg(S)$.

For the uniqueness, it is enough to remark that if $(Q,R) \neq (0,0)$ then $\exp(QP + R) \neq 0$, so $S \neq 0$.

Definition 1.2 *Let I be a left ideal of $\mathcal{D}$; we define a subset of $\mathbf{N}^2$:*

$$\text{Exp}(I) = \{\exp(P) \quad ; \quad P \in I\} \quad .$$

The fact that I is an ideal implies

$$\text{Exp}(I) = \text{Exp}(I) + \mathbf{N}^2 \quad .$$

We define

$$p = \min\{\deg(P) \quad \text{such} \quad \text{that} \quad P \in I\}$$

and for $j \geq p$

$$\alpha_j = \min\{\text{val}(P) \quad \text{such} \quad \text{that} \quad P \in I \quad \text{and} \quad \deg(P) = j\} \quad .$$

From the equality $\text{Exp}(I) = \text{Exp}(I) + \mathbf{N}^2$, we deduce that $\alpha_j \geq \alpha_{j+1}$; let

$$q = \min\{j \quad ; \quad \alpha_j = \alpha_{j+1} = \alpha_{j+2} = \ldots\} \quad .$$

The boundary of $\text{Exp}(I)$ is characterised by

$$\text{Es}(I) = \{(\alpha_p, p), (\alpha_{p+1}, p+1), \ldots, (\alpha_q, q)\} \quad .$$

We call this set the stairs of I.

Example

$$I = \mathcal{D}(x\partial_x - 2)^2 + \mathcal{D}\partial_x^3(x\partial_x - 2) ,$$
$$\text{Es}(I) = \{(2,2), (2,3), (1,4)\} \quad .$$

Definition 1.3 *Let I be a left ideal of $\mathcal{D}$; let*

$$\text{Es}(I) = \{(\alpha_p, p), (\alpha_{p+1}, p+1), \ldots, (\alpha_q, q)\}$$

be the stairs of I; The data $\{P_p, P_{p+1}, \ldots, P_q\}$ such that

$$P_k \in I, \quad k \in \{p, \ldots, q\} \quad \text{and} \quad \exp(P_k) = (\alpha_k, k),$$

is called a standard basis of I.

Theorem 1.1 *(Division theorem by an ideal I)*
Let I be a left ideal of $\mathcal{D}$, and $\{P_p, P_{p+1}, \ldots, P_q\}$ a standard basis of I. For all P in $\mathcal{D}$, there exist

$$(f_p, \ldots, f_{q-1}) \in \mathbb{C}\{x\}^{q-p} \quad and \quad Q_q \in \mathcal{D} \quad , \quad R \in \mathcal{D},$$

such that

$$P = f_p P_p + f_{p+1} P_{p+1} + \ldots + f_{q-1} P_{q-1} + Q_q P_q + R$$

where

$$R = \sum_{l=p}^{\deg(P)} \sum_{k=0}^{\alpha_l - 1} r_{k,l} x^k \partial_x^l + S \quad and \quad \deg(S) < p \quad , \quad r_{k,l} \in \mathbb{C} \quad .$$

Moreover this data is unique and $P \in I$ if and only if $R = 0$.

Proof : We begin by dividing A by P_q, then we divide the remainder of this first division by P_{q-1}, etc ... The uniqueness is obvious. Now if P is also in I, R is in I but by definition of $\exp(I)$, this may happen only if $R = 0$.

Corollary 1.2 *A standard basis of I is a system of generators of I.*

Let $\{P_p, P_{p+1}, \ldots, P_q\}$ be a standard basis of I. We can assume that the elements P_k of the standard basis are normalized (by a multiplication by a unit of D):

$$P_k = x^{\alpha_k} \partial_x^k + \sum_{i=0}^{k-1} a_i \partial_x^i \quad .$$

We have

$$\deg(P_j - x^{\alpha_j} \partial_x^j) \le j \quad for \quad j = p, \ldots, q \quad .$$

By division of $x^{\alpha_j - 1 - \alpha_j} P_j$ by $\{P_{j-1}, \ldots, P_p\}$ we obtain the relations

$$\mathcal{R}_j : x^{\alpha_j - 1 - \alpha_j} P_j = (\partial_x + u_{j,j-1}) P_{j-1} + u_{j,j-2} P_{j-2} + \ldots + u_{j,p} P_p$$

where $u_{j,k} \in \mathbb{C}\{x\}$ for all $p + 1 \le j \le q$. We can show [B.M.] that these relations form a basis of the relations between the elements $\{P_p, P_{p+1}, \ldots, P_q\}$, so we get the following free resolution of I as a left $\mathcal{D}$ module:

$$0 \longrightarrow \mathcal{D}^{q-p} \overset{\mathcal{R}}{\longrightarrow} \mathcal{D}^{q-p-1} \longrightarrow I \longrightarrow 0 \quad .$$

Lemma 1.3 *For any two operators P and Q such that $\deg(P) = 0$ and $\mathrm{val}(Q) = 0$, we have $\mathcal{D} = \mathcal{D}P + \mathcal{D}Q$.*

Proof : Let $P = x^m + \ldots \in \mathbf{C}\{x\}$, $Q = \partial_x^q + \ldots$.

We consider the commutator of P and Q:

$$Q_1 = [P, Q] = PQ - QP = \partial_x(p)\partial_x^{q-1} + \ldots$$

Next, we consider

$$Q_2 = \partial_x Q_1 - \partial_x(p)Q = \partial_x^2(p)\partial_x^{q-1} + \ldots \ ;$$

the valuation of the Q_j decreases so we find in the ideal generated by P and Q an operator Q_m whose exponent is $(0, q-1)$. By iteration on q, we now find an operator Q' in $\mathcal{D}P + \mathcal{D}Q$ whose exponent is $(0,0)$, which is a unit of $\mathcal{D}$.

Lemma 1.4 *Let I be a left ideal of $\mathcal{D}$, and let $\{P_p, P_{p+1}, \ldots, P_q\}$ be a standard basis of I.*

   1. $\forall k \ \ \exists l$ such that $x^l P_k \in \mathcal{D}P_p$;

   2. $\forall k \ \ \exists U \in \mathcal{D}$ such that $\mathrm{val}(U) = 0$ and $U F_k \in \mathcal{D}F_q$.

Proof : We deduce this lemma from the relations $\mathcal{R}_j$ between the elements of the standard basis $\{P_p, P_{p+1}, \ldots, P_q J\}$.

This lemma implies the following proposition:

Proposition 1.2 *Let I be a left ideal of $\mathcal{D}$, $\{P_p, P_{p+1}, \ldots, P_q\}$ a standard basis of I. Then I is generated by P_p and P_q:*

$$I = \mathcal{D}P_p + \mathcal{D}P_q \ .$$

If $\mathrm{val}P_q = \deg P_q = 0$, then $I = \mathcal{D}$.

1.3. Solutions of a differential system

Let $\mathcal{F}$ be a left $\mathcal{D}$ module "viewed" as a space of "generalized" functions. For $f \in \mathcal{F}$ and $P \in \mathcal{D}$ we denote $P.f$ by $P(f)$. Let I be a left ideal of $\mathcal{D}$.

Definition 1.4 *$Sol(I, \mathcal{F})$ will denote the vector space*

$$Sol(I; \mathcal{F}) = \{u \in \mathcal{F} \ \ \text{such that} \ \ \forall P \in I \ , \ \ P(u) = 0\} \ .$$

Definition 1.5 *Let* $P_1, \ldots, P_k$ *be such that* $P_i \in \mathcal{D}$. *We denote by* $Sol(P_1, \ldots, P_k; \mathcal{F})$ *the vector space*

$$Sol(P_1, \ldots, P_k; \mathcal{F}) = \{u \in \mathcal{F} \quad \text{such that} \quad P_1(u) = \ldots = P_k(u) = 0\}.$$

We have shown that $\mathcal{D}$ is noetherian. Each ideal I is generated by a finite number of generators, $P_1, \cdots, P_k$; so

$$Sol(I; \mathcal{F}) = Sol(P_1, \ldots, P_k; \mathcal{F}).$$

Let ϵ be a strictly positive real number, let $D_\epsilon = \{x \in \mathbf{C} \quad ; \quad |x| < \epsilon\}$; we denote by $\tilde{\mathcal{O}}(D_\epsilon - \mathbf{R}^+)$ the analytic functions on $D_\epsilon - \mathbf{R}^+$ with multivalued continuation. And we let:

$$\tilde{A} = \lim_{\vec{\epsilon}} \tilde{\mathcal{O}}(D_\epsilon - \mathbf{R}^+)$$

In this subsection, we consider the following function spaces $\mathcal{F}$:

1. $\mathbf{C}[[x]]$

2. $\mathbf{C}\{x\}$

3. $\tilde{A}$

4. $\tilde{A}/\mathbf{C}\{x\}$

5. $\tilde{N} \subset \tilde{A}$ the set of Nilsson functions:

$$f = \sum_\alpha (\sum_{i=0}^{m_\alpha} f_{\alpha,i}(x) Log^i x) x^\alpha$$

where α is in a finite set of $\mathbf{C}$ and $f_{\alpha,i}$ is an analytic function.

1.3.1. $Sol(I; \mathcal{F})$

Let I be a left ideal of $\mathcal{D}$; we denote by $(P_1, \ldots, P_p)$ a standard basis of I and $Es(I) = (p, \alpha_p), \ldots, (q, \alpha_q)$ its stairs.

Lemma 1.5 *If $\mathcal{F}$ is a $\mathcal{D}$ module without $\mathbf{C}\{x\}$ torsion (for example: $\mathbf{C}\{x\}$, $\mathbf{C}[[x]]$, $\tilde{A}$). Then $Sol(I; \mathcal{F})$ is equal to $Sol(P_p; \mathcal{F})$, the set of solutions of P_p taking values in $\mathcal{F}$.*

Proof :   It is clear that

$$Sol(I; \mathcal{F}) \subset Sol(P_p; \mathcal{F}) \quad .$$

Suppose $u \in \mathcal{F}$ and $P_p(u) = 0$; by lemma 1.4

$$\forall k \in \{1, \ldots, p\} \quad \exists l \in \mathbf{N} \quad ; \quad x^l P_k \in \mathcal{D} P_p \quad ,$$

then $x^l P_k(u) = 0$, thus $P_k(u) = 0$. As $\{P_i\}_{i \in \{1,\ldots,p\}}$ are generators of I, we are done.

Now, consider the solutions in $\tilde{A}$, ; Cauchy's theorem implies

$$\dim(Sol(P_p; \tilde{A})) = p$$

and we obtain:

Proposition 1.3
$$Sol(I; \tilde{A}) = Sol(F_p; \tilde{A}) \ ,$$
$$\dim(Sol(I; \mathcal{F})) = p \quad .$$

To compute the solutions of I in $\tilde{A}/\mathbf{C}\{x\}$. We begin by determining the dimension of $Sol(P; \tilde{A}/\mathbf{C}\{x\})$. For this purpose, we consider the following commutative diagram:

$$
\begin{array}{ccccccccc}
0 & \rightarrow & \mathbf{C}\{x\} & \rightarrow & \tilde{A} & \rightarrow & \tilde{A}/\mathbf{C}\{x\} & \rightarrow & 0 \\
& & P \downarrow & & P \downarrow & & P \downarrow & & \\
0 & \rightarrow & \mathbf{C}\{x\} & \rightarrow & \tilde{A} & \rightarrow & \tilde{A}/\mathbf{C}\{x\} & \rightarrow & 0
\end{array}
$$

The horizontal sequences are exact. By an index theorem of Malgrange [M 1] the operator

$$P : \mathbf{C}\{x\} \rightarrow \mathbf{C}\{x\}, \quad f \mapsto P(f),$$

has an index equal to

$$\dim(\ker P) - \dim(\mathrm{coker} P) = \deg(P) - \mathrm{val}(P).$$

On the other hand, Cauchy's theorem implies that

$$P : \tilde{A} \rightarrow \tilde{A}, \quad f \mapsto P(f),$$

is surjective and its kernel is a space of dimension $\deg(P)$. From the long exact sequence associated to our diagram, we obtain

$$\dim Sol(P; \tilde{A}/\mathbf{C}\{x\}) = \mathrm{val}(P) \quad .$$

Proposition 1.4

$$Sol(I; \tilde{A}/\mathbf{C}\{x\}) = Sol(F_q; \tilde{A}/\mathbf{C}\{x\}) \ ,$$

$$\dim Sol(I; \tilde{A}/\mathbf{C}\{x\}) = \alpha_q \quad .$$

Proof :   The proof is similar to the proof of the previous proposition. It will follow from the inclusion

$$Sol(F_q; \tilde{A}/\mathbf{C}\{x\}) \subset Sol(I; \tilde{A}/\mathbf{C}\{x\}) \quad .$$

By lemma 1.4 $\forall k$ $\exists U \in \mathcal{D}$ such that $\mathrm{val}(U) = 0$ and $UF_k \in \mathcal{D}F_q$. Assume that $f \in \tilde{A}$ and $g \in \mathbf{C}\{x\}$ such that $F_q(f) = g$; we conclude that

$$U(F_k(f)) \in \mathbf{C}\{x\} \quad .$$

But $\mathrm{val}(U)$ is zero, so by Cauchy's theorem we have $F_k(f) \in \mathbf{C}\{x\}$. This proves our inclusion.

Proposition 1.5   *1. If P vanishes on $Sol(I; \tilde{A})$ there exists $k \in \mathbf{N}$ such that $x^k P \in \mathcal{D}F_p$.*

  2. If P vanishes on $Sol(I; \tilde{A}/\mathbf{C}\{x\})$, then there exists $U \in \mathcal{D}$ with $\mathrm{val}(U)$ zero, such that $UP \in \mathcal{D}F_q$.

  3. If P vanishes both on $Sol(I; \tilde{A})$ and on $Sol(I; \tilde{A}/\mathbf{C}\{x\})$, then $P \in I$.

Proof :   Using the division theorem by the operator F_p,

$$\exists k \in \mathbf{N} \quad x^k P = AF_p + R \quad \text{with} \quad \deg(R) < p \quad .$$

But R vanishes on $Sol(I; \tilde{A})$ and $\deg(R) < \dim(Sol(I; \tilde{A}))$. Thus R is zero by Cauchy's theorem. So we get 1.
For 2: there is a proof of this result in [Mai] (Proposition 1.2.5, page 107).
By 1 and 2 of this proposition, we can find x^k and U with $\mathrm{val}(U)$ equal to zero such that $x^k P$ and UP are in I. But by Lemma 1.3, there exist two operators A and B such that

$$1 = Ax^k + BU \quad .$$

Therefore $P = Ax^k P + BUP$ is in I.

1.3.2. Regularity and irregularity

Let

$$P = \sum_{k=0}^{\deg P} a_k(x)\partial^k \quad , \quad \text{where} \quad a_k(x) = \sum_{j=0}^{\infty} a_{k,j}x^j \quad .$$

We denote by $p(P)$ the weight of P and by $in(P)$ the initial part of P:

$$p(P) = sup\{k - \text{val}(a_k) \; ; \; a_k \neq 0\} \quad , \quad in(P) = \sum_{k-l=p(P)} a_{k,l}x^l\partial^k.$$

We have the following properties:

- $p(P_1 P_2) = p(P_1) + p(P_2)$

- $in(P_1 P_2) = in(P_1).in(P_2)$

- if $p(P_1) < p(P_2)$ then $p(P_1 + P_2) = p(P_2)$ and $in(P_1 + P_2) = in(P_2)$

- if $p(P_1) = p(P_2)$ and $in(P_1 + P_2) \neq 0$ then $p(P_1 + P_2) = p(P_1)$ and $in(P_1 + P_2) = in(P_2)$

- if $p(P_1) = p(P_2)$ and $in(P_1 + P_2) = 0$ then $p(P_1 + P_2) < p(P_1)$

Following B. Malgrange and M. Kashiwara this filtration of $\mathcal{D}$ by the weights is called the V filtration.

Definition 1.6 *An operator P in $\mathcal{D}$ is called regular if P and $in(P)$ have the same degree. We can write this condition*

$$p(P) = \deg(P) - \text{val}(P) \quad .$$

We deduce from the above formulas that PQ is regular if and only if P and Q are regular.

Proposition 1.6 *For a differential operator P, the following three properties are equivalent:*

1. *P is regular ;*

2. *$Sol(P; \tilde{A})$ consists of functions in the Nilsson classes ;*

3. *$Sol(P; \tilde{A}/\mathbb{C}\{x\})$ consists of functions in the Nilsson classes.*

Proof : $1 \Rightarrow 2$ is a classical result, see [W].

$2 \Rightarrow 3$ is obtained by variation of constants. The determinant of a fundamental system of solutions of $P(u) = 0$ is of the form $x^\alpha g(x)$ where $g(x)$ is in $C\{x\}$.

$3 \Rightarrow 1$: let $\dot{f}_1, \cdots, \dot{f}_r$ be a basis of $Sol(P; \tilde{A}/C\{x\})$. We can construct a regular operator R such that $R(\dot{f}_i) = 0$ for all i. So R vanishes on $Sol(P; \tilde{A}/C\{x\})$; by Proposition 1.5 there exists an operator U with $val(U) = 0$ and an operator A such that: $UR = AP$. As U and R are regular, P is also regular.

Definition 1.7 *([M 1]) We define the irregularity of a differential operator P to be the integer*

$$i(P) = p(P) - (\deg(P) - val(P)) \quad .$$

By B. Malgrange's result [M 1] the linear application:

$$P : C\{x\} \to C\{x\} , \qquad f \mapsto P(f) ,$$

has an index, $\deg(P) - val(P)$, called the convergent index of P.
On the other hand, the linear application

$$P : C[[x]] \to C[[x]] , \qquad f \mapsto P(f) ,$$

has an index $p(P)$, called the formal index of P.

We can reformulate the previous definition:

Corollary 1.6 *[M 1] The irregularity of a differential operator P is the difference between the formal index of P and the convergent index of P. An operator P is regular if its irregularity is zero.*

Lemma 1.7 *Let I be a left ideal of $\mathcal{D}$; then all the elements of a standard basis of I have the same irregularity*

Proof :   Let $(F_p, \ldots, F_q)$ be a standard basis of I. Recall the relations

$$\mathcal{R}_{j+1} : x^{\alpha_j - \alpha_{j+1}} F_{j+1} = (D + u_{j,j}) F_j + u_{j,j-1} F_{j-1} + \ldots + u_{j,p} F_p$$

where $u_{j,k} \in C\{x\}$ for all $p + 1 \le j \le q$.
By definition of j and α_j,

$$\deg(F_{j+1}) - val(F_{j+1}) = j + 1 - \alpha_{j+1} = \deg(F_j) - val(F_j) + \alpha_j - \alpha_{j+1} + 1 \quad .$$

To show that $i(F_{j+1}) = i(F_j)$, we prove by induction that

$$p(F_{j+1}) = p(F_j) + \alpha_j - \alpha_{j+1} + 1 \quad .$$

We use the relations $\mathcal{R}_{j+1}$ and we proceed by induction: for $k < j$,

$$p(u_{j,k}F_k) < p(F_j) \quad ;$$

on the other hand

$$p((D + u_{j,j})F_j) = p(F_j) + 1 \quad .$$

We conclude by using the formulas for the weight p.

Definition 1.8 *We define the irregularity $i(I)$ of a left ideal I of $\mathcal{D}$ to be the integer*

$$i(I) = inf\{i(P) \; ; \; P \in I - \{0\}\} \quad .$$

Proposition 1.7 *If $(F_p, \ldots, F_q)$ is a standard basis of I, we have*

$$i(I) = i(F_p) = i(F_{p+1}) = \ldots = i(F_p) \quad .$$

Proof : By Lemma 1.7, it is enough to prove that $i(I) = i(F_p)$. In fact, it is enough to prove that if $P \in I$, $i(P) \geq i(F_p)$. Let $P \in I$, then by Proposition 1.5:

$$\exists Q \in \mathcal{D} \quad , \quad \exists \alpha \in \mathbf{N} \quad \text{such} \quad \text{that} \quad x^\alpha P = QF_p \quad .$$

But $i(x^\alpha) = 0$, so by the additivity of the irregularity

$$i(P) = i(Q) + i(F_p) \geq i(F_p) \quad .$$

Definition 1.9 *A left ideal I of $\mathcal{D}$ is regular if $i(I) = 0$.*

The preceding results show the following proposition.

Proposition 1.8 *Let I be a non-trivial left ideal of $\mathcal{D}$, and let $(F_p, \ldots, F_q)$ be a standard basis of I. The following assertions are equivalent:*

1. *I is regular.*

2. *An element F_k of the standard basis is regular.*

3. *$Sol(I; \tilde{A})$ consists of functions of Nilsson classes.*

4. *$Sol(I; \tilde{A}/\mathbf{C}\{x\})$ consists of functions of Nilsson classes.*

5. *$\exists P \in I$ such that P is a regular operator.*

1.3.3.   The pair E $\rightleftarrows$ F associated to an ideal

We consider the canonical surjection u:

$$\tilde{A} \overset{u}{\to} \tilde{A}/\mathbf{C}\{x\}$$

which associates to each element its class in the quotient. It is a morphism of $\mathcal{D}$ modules. For a left ideal I of $\mathcal{D}$, u induces a map:

$$u : Sol(I,\tilde{A}) \to Sol(I,\tilde{A}/\mathbf{C}\{x\})   .$$

We now consider the map $T : \tilde{A} \to \tilde{A}$, $f \mapsto T(f)$; where $T(f)$ is obtained by taking the prolongation of f along a path. T is an isomorphism of left $\mathcal{D}$ modules. The restriction of T to $\mathbf{C}\{x\}$ is the identity. Thus, we can define a morphism of left $\mathcal{D}$ modules:

$$\text{var} :   \tilde{A}/\mathbf{C}\{x\} \longrightarrow \tilde{A}   ,   f \mapsto Tf - f .$$

We have

$$\text{var} \circ u + Id = T   .$$

For any left ideal I of $\mathcal{D}$, this morphism var induces the following morphism of C vector spaces:

$$\text{var} :   Sol(I,\tilde{A}/\mathbf{C}\{x\}) \to Sol(I,\tilde{A}) ,   f \mapsto \text{var}(f) .$$

We introduce the category Θ whose elements are the pairs E $\rightleftarrows$ F where E and F are C vector spaces of finite dimension, u and v are linear applications of C vector spaces and $v \circ u + Id$ is a isomorphism; and where the morphisms between two objects are the obvious ones. We can show that Θ is an additive and abelian category. So any object of Θ is a direct sum of indecomposable objects.

Definition 1.10 *To a left ideal I of $\mathcal{D}$ we associate the pair $E(I) \rightleftarrows F(I)$ formed by*

$$Sol(I,\tilde{A}) \overset{u}{\underset{\text{var}}{\rightleftarrows}} Sol(I,\tilde{A}/\mathbf{C}\{x\})   .$$

We recall that if I is a non-trivial left ideal of $\mathcal{D}$, and if I is regular, then

$$Sol(I,\tilde{A}) \overset{u}{\underset{\text{var}}{\rightleftarrows}} Sol(I,\tilde{A}/\mathbf{C}\{x\})$$

is a pair consisting of functions of Nilsson classes.
In [B.M.], we prove by a direct computation

Proposition 1.9 *Let I be a non-trivial regular left ideal of $\mathcal{D}$.*

$$\oplus \quad (E_{\lambda,l} \rightleftarrows F_{\lambda,l})$$

is a decomposition in Θ of

$$Sol(I, \tilde{A}) \xleftarrow[\overrightarrow{var}]{u} Sol(I, \tilde{A}/\mathbf{C}\{x\})$$

into a direct sum of indecomposable objects.
If

$$I_{\lambda,l} = \{P \in \mathcal{D} ; \; \forall f \in (E_{\lambda,l}) \quad and \quad \forall g \in (F_{\lambda,l}) \; : \; P(f) = 0 \; P(g) = 0\} \quad ,$$

then :

1. $I = \cap I_{\lambda,l}$

2. $\mathcal{D}/I \simeq \oplus(\mathcal{D}/I_{\lambda,l})$

3. $Sol(I_{\lambda,l}, \tilde{A}) = E_{\lambda,l}$ *and* $Sol(I_{\lambda,l}, \tilde{A}/\mathbf{C}\{x\}) = F_{\lambda,l}$

Moreover any quotient $\mathcal{D}/I_{\lambda,l}$ is isomorphic as $\mathcal{D}$ module to a unique module in the following list:

$$\mathcal{D}/\mathcal{D}(x\partial - \alpha)^r \quad \mathcal{D}/\mathcal{D}(\partial x)^r \quad \mathcal{D}/\mathcal{D}(x\partial)^r \quad \mathcal{D}/\mathcal{D}(\partial x)^r \partial \quad \mathcal{D}/\mathcal{D}(x\partial)^r x$$

where $r \in \mathbf{N} - \{0\}$ and $\alpha \in \mathbf{C} - \mathbf{Z}$.

1.4.   Holonomic $\mathcal{D}$ modules in one variable

First we recall some general results on finite left $\mathcal{D}$ module in n variables.
We will use the following result given in the lectures of B. Malgrange (this book). Any finite $\mathcal{D}$ module has a dimension $\leq n$, a left $\mathcal{D}$ module is said to be holonomic if its dimension is n, so its characteristic variety is a lagrangian variety. We preserve his notations. Let M_k be a good filtration of M; then $gr(M)$ is a $gr(\mathcal{D}) = \mathcal{O}[\xi]$ module. Let $\xi \in \mathbf{C}^n$, in [L.M] [G.M.] the following result is shown

$$\dim_{\mathbf{C}} gr(M)/(x,\xi)^k gr(M)$$

is a polynomial in k for k large enough. This polynomial can be written as

$$\frac{m(M)}{\dim(M)!} k^{\dim(M)} + a_{k-1} k^{\dim-1} + \dots$$

where $\dim(M)$ is the dimension of M, $m(M)$ is a positive integer independent of the good filtration, which we call the multiplicity of M. If

$$0 \to M' \to M \to M'' \to 0$$

is an exact sequence of finite $\mathcal{D}$ modules, we have

- $\dim(M) = \max\{\dim(M'), \dim(M'')\}$.

- If $\dim(M') < \dim(M'')$ then $m(M) = m(M'')$.

- If $\dim(M') < \dim(M'')$ then $m(M) = m(M')$.

- If $\dim(M') = \dim(M'')$ then $m(M) = m(M') + m(M'')$.

Any finite $\mathcal{D}$ module has a dimension $\leq n$; then it is a consequence of the previous properties that if M' is a strict submodule of a holonomic $\mathcal{D}$ module M, then M' is holonomic and $m(M') < m(M)$.

Proposition 1.10 *Let M be a holonomic left $\mathcal{D}$ module in $\underline{n}$ of variables: then M is of finite length (any increasing sequence of $\mathcal{D}$ submodules of M is stationary).*

It is not very difficult to prove that the ring $\mathcal{D}$ of the differential operators in $\underline{n}$ variables is simple. That is, if J is a 2 sided ideal, then $J = 0$ or $J = \mathcal{D}$. However $\mathcal{D}$ has infinite length as a left $\mathcal{D}$ module: for example the sequence $(\partial_1, \ldots, \partial_n)^k \mathcal{D}$ is not stationary.
We now recall a general result of Stafford [Bjo] whose proof is elementary:

Proposition 1.11 *Let A be a simple ring that is infinite as a left A module. Then every left A module of finite length is cyclic, i.e. it can be generated by one element.*

Corollary 1.8 *Let M be a holonomic left $\mathcal{D}$ module in $\underline{n}$ variables. Then M is cyclic: i.e. there exists a left ideal I of $\mathcal{D}$ such that $M \simeq \mathcal{D}/I$ as $\mathcal{D}$ modules.*

The characteristic variety of a $\mathcal{D}$ holonomic module is a germ of a lagrangian variety. Let $\mathrm{char}(M) = \cup V_\alpha$ be the decomposition into irreducible components. V_α is defined from the annihilating ideal $\overline{\mathcal{I}}$ of $\mathrm{gr}(M)$ for a any good filtration of M. $\overline{\mathcal{I}}$ defines an ideal $\overline{\mathcal{I}}_a$ of $\mathbf{C}\{x - a\}[\xi]$ for a a point in the neighborhood of 0. Let $(a, \underline{\xi})$ be a generic point of V_α; as for $m(M)$, when k is large enough,

$$\dim_{\mathbf{C}} \frac{\mathbf{C}\{x - a\}[\xi]}{(x - a, \xi - \underline{\xi})^k + \overline{\mathcal{I}}_a} = \frac{m_\alpha}{n!} k^n + \ldots$$

is a polynomial of degree n and m_α is an integer. Moreover m_α is independent of the good filtration. We call m_α the multiplicity of V_α in $\mathrm{char}(M)$, where

Definition 1.11 $\mathrm{Char}(M) = \sum m_\alpha V_\alpha$ *is the characteristic cycle of the holonomic* $\mathcal{D}$ *module* M.

We now go back to the <u>one</u> variable case.

Let I be a non-zero left ideal of $\mathcal{D}$, and let $(F_p, \ldots, F_q)$ be a standard basis of I. The natural filtration $\mathcal{D}_k$ of $\mathcal{D}$ (paper of B. Malgrange section 2) induces a natural filtration of I,

$$I_k = \mathcal{D}_k \cap I ,$$

and we have a natural injection:

$$\mathrm{gr}(I) = \oplus_{k \in \mathbf{N}} \frac{\mathcal{D}_k \cap I}{\mathcal{D}_{k-1} \cap I} \to \mathrm{gr}(\mathcal{D}) = \oplus_{k \in \mathbf{N}} \frac{\mathcal{D}_k}{\mathcal{D}_{k-1}} \simeq \mathbf{C}\{x\}[\xi] \quad .$$

Then $\mathrm{gr}(I)$ is identified with the ideal in $\mathbf{C}\{x\}[\xi]$ consisting of the principal symbols of operators of I. By the division theorem 1.1, this ideal is

$$(x^{\alpha_p}\xi^p, \ldots, x^{\alpha_q}\xi^q)\mathbf{C}\{x\}[\xi] \quad .$$

Let us consider the left $\mathcal{D}$ module $\mathcal{D}/I$. The natural filtration of $\mathcal{D}$ induces a natural filtration in $\mathcal{D}/I$

$$\left(\frac{\mathcal{D}}{I}\right)_k = \frac{\mathcal{D}_k + I}{I}$$

and we have

$$\mathrm{gr}\left(\frac{\mathcal{D}}{I}\right) \simeq \frac{\mathrm{gr}(\mathcal{D})}{\mathrm{gr}(I)} \simeq \frac{\mathbf{C}\{x\}[\xi]}{(x^{\alpha_p}\xi^p, \ldots, x^{\alpha_q}\xi^q)\mathbf{C}\{x\}[\xi]} \quad .$$

We compute the characteristic variety of $\mathbf{D}/I$.

If $I \neq \mathcal{D}$, we have seen that $(\alpha_q, p) \neq (0,0)$, and then we have:

- If $\alpha_q \neq 0$ and $p \neq 0$, then char $(\mathcal{D}/I) = \mathbf{C}^n \cup T_0^*\mathbf{C}^n$.

- If $\alpha_q = 0$ and $p \neq 0$, then char$(\mathcal{D}/I) = \mathbf{C}^n$.

- If $\alpha_q \neq 0$ and $p = 0$, then char$(\mathcal{D}/I) = T_0^*\mathbf{C}^n$.

We compute the multiplicity of M.

If $I \neq \mathcal{D}$, $\alpha_q + p$ is not zero. For k large enough,

$$\dim \frac{\mathrm{gr}(\mathcal{D}/I)}{(x,\xi)^k \mathrm{gr}(\mathcal{D}/I)} = \dim \frac{\mathbf{C}\{x\}[\xi]}{\mathrm{gr}(I) + (x,\xi)^k} = (\alpha_q + p)k + const \quad .$$

The multiplicity of M is $p + \alpha_q$. Analogous calculations show that the multiplicity of char$(\mathcal{D}/I)$ in $\mathbf{C}^n$ is p and the multiplicity of $T_0^*\mathbf{C}^n$ is α_q.

So, the characteristic cycles of $\mathcal{D}/I$ are

$$\mathrm{Char}(M) = p\mathbf{C}^n + \alpha_q T_0^*\mathbf{C}^n$$

Hence, we have proved the proposition:

Proposition 1.12 *Let M be a left $\mathcal{D}$ module. Then M is holonomic if and only if there is a left ideal I of $\mathcal{D}$ such that*

$$I \neq 0 \quad , \quad I \neq \mathcal{D} \quad , \quad \text{and} \quad M \simeq \frac{\mathcal{D}}{I}.$$

Moreover if $(p, \alpha_p), \ldots, (q, \alpha_q)$ is the stair of I the Characteritic cycle of M is

$$\text{Char}(M) = p\mathbf{C}^n + \alpha_q T_0^* \mathbf{C}^n \quad .$$

Consider now a left holonomic $\mathcal{D}$ module M and a left ideal I of $\mathcal{D}$ such that $M \simeq \mathcal{D}/I$. Let $\mathcal{F}$ be a space of functions (subsection 1.3.). The solution of M in $\mathcal{F}$ is the space $Hom_{\mathcal{D}}(M, \mathcal{F})$. As a vector space it is isomorphic to $Sol(I, \mathcal{F})$.

Using homological algebra we define the complex of vector spaces $RHom_{\mathcal{D}}(M, \mathcal{F})$. We can compute this complex with a $\mathcal{D}$-free resolution of M or a $\mathcal{D}$-injective resolution of $\mathcal{F}$. Let

$$\ldots \overset{d_{-2}}{\to} F_1 \overset{d_{-1}}{\to} F_0 \overset{d_0}{\to} F_1 \overset{d_1}{\to} \ldots \overset{d_{k-1}}{\to} F_k \overset{d_k}{\to} \ldots$$

be a complex which determines $RHom_{\mathcal{D}}(M, \mathcal{F})$. We denote $Ext_{\mathcal{D}}^k(M, \mathcal{F}) = ker(d_k)/im(d_{k-1})$, the cohomology groups of this complex which are independent of the resolution. If these vector spaces are of finite dimension, we define the Euler characteristic:

$$\chi(RHom_{\mathcal{D}}(M, \mathcal{F})) = \sum_i (-1)^i \dim_{\mathbf{C}} Ext_{\mathcal{D}}^i(M, \mathcal{F}) \quad .$$

Example: If $M = \mathcal{D}/\mathcal{D}P$ where P is a differential operator we have the following free resolution of M:

$$\begin{array}{ccccccccc} 0 & \to & \mathcal{D} & \to & \mathcal{D} & \overset{\pi}{\to} & \mathcal{D}/\mathcal{D}P & \to & 0 \\ & & Q & \mapsto & QP & & & & \end{array}$$

Then $RHom_{\mathcal{D}}(\mathcal{D}/\mathcal{D}P, \mathcal{F})$ is the complex of $\mathbf{C}$ vector spaces

$$\begin{array}{ccccccc} 0 & \to & Hom_{\mathcal{D}}(\mathcal{D}, \mathcal{F}) & \to & Hom_{\mathcal{D}}(\mathcal{D}, \mathcal{F}) & \to & 0 \\ & & \overset{\simeq}{\mathcal{F}} & \to & \overset{\simeq}{\mathcal{F}} & & \\ & & f & \mapsto & P(f) & & \end{array}$$

In subsection 1.2., we have shown how to compute a free resolution of $M = \mathcal{D}/I$. Now we take $\mathcal{F} = \mathbf{C}\{x\}$, and consider the exact sequence of $\mathcal{D}$ modules

$$0 \to \mathbf{C}\{x\} \to \tilde{A} \overset{can}{\to} \tilde{A}/\mathbf{C}\{x\} \to 0 \quad .$$

By the results of subsection 1.3., it is not too difficult to prove that $\tilde{A}$ and $\tilde{A}/C\{x\}$ are injective $\mathcal{D}$ modules. Hence, the complex $vRHom_{\mathcal{D}}(M, C\{x\})$ is

$$0 \to Hom_{\mathcal{D}}(M, \tilde{A}) \overset{can}{\to} Hom_{\mathcal{D}}(M, \tilde{A}/C\{x\}) \to 0$$

This complex is isomorphic to

$$0 \to Hom_{\mathcal{D}}(\mathcal{D}/I, \tilde{A}) \overset{can}{\to} Hom_{\mathcal{D}}(\mathcal{D}/I, \tilde{A}/C\{x\}) \to 0$$
$$\cong \qquad\qquad\qquad\qquad \cong$$
$$0 \to \qquad E(I) \qquad \overset{can}{\to} \qquad\qquad F(I) \qquad\qquad \to 0$$

Hence we have the following proposition:

Proposition 1.13 *Let M be a left holonomic $\mathcal{D}$ module in one variable with characteristic cycle $p\mathbf{C}^n + \alpha T_0^* \mathbf{C}^n$. Then*

1. $Ext_{\mathcal{D}}^i(M, C\{x\}) = 0$ *for* $i \notin \{0, 1\}$.

2. $Ext_{\mathcal{D}}^i(M, C\{x\})$ *are $\mathbf{C}$ vector spaces of finite dimension.*

3. $\chi(RHom_{\mathcal{D}}(M, C\{x\}) = \sum_i (-1)^i \dim_{\mathbf{C}}(Ext_{\mathcal{D}}^i(M, C\{x\})) = p - \alpha$.

Remark: This proposition has been generalized for holonomic $\mathcal{D}$ modules in several variables. The generalization of 2. is the constructibility theorem of Kashiwara [K][M.N.], the generalization of 3. is the index formula of Kashiwara [K 1].

We return to the one variable case. We can prove that if M is holonomic then $Ext_{\mathcal{D}}^i(M, C[[x]])$ are $\mathbf{C}$ vector spaces of finite dimension [M 4][Mai].

Definition 1.12 *We refer as the irregularity of M to the integer*

$$i(M) = \chi(RHom_{\mathcal{D}}(M, C[[x]])) - \chi(RHom_{\mathcal{D}}(M, C\{x\})) \quad .$$

Let I be a left ideal associated to M, we can prove that $i(M)$ is equal to the irregularity of I (subsubsection 1.3.2.). M is said to be regular if $i(M)$ is zero.

Proposition 1.14 *Let M be a left holonomic regular $\mathcal{D}$ module, M is $\mathcal{D}$ isomorphic in a unique way to a finite direct sum of the following regular holonomic $\mathcal{D}$ modules:*

$$\mathcal{D}/\mathcal{D}(x\partial - \alpha)^r ; \quad \mathcal{D}/\mathcal{D}(\partial x)^r ; \quad \mathcal{D}/\mathcal{D}(x\partial)^r ; \quad \mathcal{D}/\mathcal{D}(\partial x)^r \partial ; \quad \mathcal{D}/\mathcal{D}(x\partial)^r x ;$$

where $r \in \mathbf{N} - \{0\}$ and $\alpha \in \mathbf{C} - \mathbf{Z}$.

For the irregular $\mathcal{D}$ modules, the classification result is more difficult. The reader can consult the book of B. Malgrange ([M 4]).

2.   Holonomic $\mathcal{D}$ modules

2.1.   Introduction

We will see that a $\mathcal{D}$ module M whose characteristic variety is very simple (the conormal to a smooth variety) can be explicitly described. Then a holonomic $\mathcal{D}$ module has a very simple structure at a generic point of its characteristic variety. Another corollary is that an endomorphism of a holonomic $\mathcal{D}$ module has a minimal polynomial. This result proves the existence of a non-zero polynomial $b_f(s)$ in s satisfying a functional equation: $b_f(s)f^s \in \mathcal{D}[s]f^{s+1}$. This polynomial is called the Bernstein polynomial of f or the b-function. As Bernstein noticed [Ber] (this was its initial aim) this polynomial allows the analytic continuation of a distribution $\int \phi f^s$. We give an algorithm for computing the Bernstein polynomial of f when f has an isolated singularity. Our result restates and completes results of [B.G.M.M.] and [M 2]. When f has an isolated singularity, we prove the existence of an operator

$$S = s^{\mu(\mu-1)} + A_1 s^{\mu(\mu-1)-1} + A_{\mu(\mu-1)} \quad \text{such} \quad \text{that} \quad S f^s = 0 \quad ,$$

where $A_i \in \mathcal{D}$, with $\deg(A_i) \leq i$ and where μ is the Milnor number of f. In fact we find an equation where the operators $A_1, \ldots, A_{\mu(\mu-1)-1}$ are in **C**. Our bound $\mu(\mu-1)$ is a new result. Kashiwara proved in [K 2] that an analogous equation exists for all f. The existence of this operator is related to the regularity of the holonomic $\mathcal{D}$ modules in several variables. We conclude by extending the basic notions of regularity and irregularity to several variables. This is an active field of research. Then we quote a theorem which compares algebraic properties (in the sense of Y. Laurent [Lau 1]) of a holonomic $\mathcal{D}$ to properties of some "solutions" of this $\mathcal{D}$ module.

2.2.   Structure of coherent $\mathcal{D}$ modules supported by a point

As in all subsections of this section, we adopt the notation of the paper of B. Malgrange (in this book). We denote by $\mathcal{M} = (x_1, \ldots, x_n)$ the maximal ideal of $\mathcal{O}$.

Definition 2.1 *Let M be a left D module; M is said to be supported by $\{0\}$ if*

$$\forall u \in M \,, \exists k \in \mathbf{N} \; ; \; \mathcal{M}^k u = 0 \quad .$$

Example: $B_{pt} = \mathcal{D}/\mathcal{D}\mathcal{M}$.
To show that B_{pt} is supported by $\{0\}$, it is enough to remark that if $P \in \mathcal{D}$ has

degree k, $\mathcal{M}^{k+1}P \in \mathcal{D}\mathcal{M}$. We denote the class of 1 by δ; and let $\mathbf{C}[\partial_1, \ldots, \partial_n]$ denote the differential operators with coefficients in $\mathbf{C}$:

$$B_{pt} = \mathcal{D}/\mathcal{D}\mathcal{M} = \mathcal{D}\delta = \mathbf{C}[\partial_1, \ldots, \partial_n]\delta \quad .$$

It is clear from the definition that $T_0^*(\mathbf{C}^n) = \{(x, \xi) \in T^*(\mathbf{C}^n) \; ; \; x = 0\}$ is contained in $\mathrm{char}(M)$. An easy computation shows that in fact

$$\mathrm{char}(M) = T_0^*(\mathbf{C}^n) \quad .$$

Proposition 2.1 *Any finite type left $\mathcal{D}$ module supported by $\{0\}$ is isomorphic to a finite direct sum of copies of B_{pt} .*

Proof :   (We follow an idea of J. Briançon.)

We begin by proving two lemmas:

Lemma 2.1 *Let $A \subset \mathbf{N}^n$ be such that $A + \mathbf{N}^n = A$. We denote by $I(A)$ the ideal of $\mathcal{O}$ generated by $\{x^\beta\}_{\beta \in A}$. $c(A)$ denotes the cardinality of $\mathbf{N}^n - A$, which we assume to be finite. Then we have the following isomorphism of $\mathcal{D}$ modules*

$$\frac{\mathcal{D}}{\mathcal{D}I(A)} \simeq B_{pt}^{c(A)} \quad .$$

Proof :   If $c(A) = 1$, there $I(A) = \mathcal{M}$ and the lemma is obvious.

Otherwise, let $\alpha \notin A$ be such that $\forall i \quad x_i x^\alpha \in Iv(A)$. We denote by $\dot{x}^\alpha$ the class of x^α in $\mathcal{D}/\mathcal{D}I(A)$ and we consider the sequence of $\mathcal{D}$ modules

$$0 \to \mathcal{D}\dot{x}^\alpha \xrightarrow{i} \frac{\mathcal{D}}{\mathcal{D}I(A)} \xrightarrow{\omega} \frac{\mathcal{D}}{\mathcal{D}I(A + \alpha)} \to 0 \quad .$$

i is the canonical injection and ω is the canonical projection. This sequence is exact. Moreover, its splits. In fact, we have a retraction r of i:

$$r : \frac{\mathcal{D}}{\mathcal{D}I(A)} \to \mathcal{D}\dot{x}^\alpha, \quad i \mapsto \frac{(-1)^\alpha}{\alpha!}\partial^\alpha \quad .$$

And we are done by induction on $c(A)$.

Lemma 2.2

$$\mathbf{C} \simeq Hom_{\mathcal{D}}(B_{pt}, B_{pt}),$$

$$\lambda \mapsto \text{multiplication by } \lambda \; .$$

Proof : The injectivity is clear; for the surjectivity, we consider $\phi \in Hom_{\mathcal{D}}(B_{pt}, B_{pt})$; ϕ is determined by $\phi(\delta)$ with the condition $\mathcal{M}\phi(\delta) = 0$. An easy computation shows that $\phi(\delta)$ is the class of an element λ of **C**. Hence, ϕ is multiplication by λ.

Now, we can prove the proposition. Let M be a left $\mathcal{D}$ module of finite type supported by $\{0\}$ and $(m_1, \ldots, m_l)$ be a system of generators of M. By hypothesis, there exists k_i in **N** such that

$$\mathcal{M}^{k_i} m_i = 0 \quad .$$

Thus, there is a surjection of $\mathcal{D}$ modules

$$\oplus \frac{\mathcal{D}}{\mathcal{D}\mathcal{M}^{k_i}} \xrightarrow{\epsilon} M \to 0 \quad .$$

But, $\ker(\epsilon)$ is again of finite type and supported by $\{0\}$. Thus, we can find another surjection:

$$\oplus_{i=1}^{p} \frac{\mathcal{D}}{\mathcal{D}\mathcal{M}^{l_i}} \xrightarrow{\psi} \ker(\epsilon) \to 0 \quad .$$

So we obtain an exact sequence:

$$\oplus_{i=1}^{p} \frac{\mathcal{D}}{\mathcal{D}\mathcal{M}^{l_i}} \to \oplus \frac{\mathcal{D}}{\mathcal{D}\mathcal{M}^{k_i}} \xrightarrow{\epsilon} M \to 0 \quad .$$

Applying Lemma 2.2, we deduce an exact sequence:

$$\oplus B_{pt} \xrightarrow{\phi} \oplus B_{pt} \to M \to 0 \quad ,$$

by Lemma 2.2, ϕ is given by a matrix with complex entries. After a linear change of coordinates, we can assume that ϕ is diagonal with 1 or 0 on the diagonal. The quotient M is thus isomorphic to a sum of copies of B_{pt}.

Remark: The following are equivalent.

- M is a left $\mathcal{D}$ module of finite type supported by $\{0\}$.

- M is a left $\mathcal{D}$ module of finite type with characteristic variety equal to $T_0^*(\mathbf{C}^n)$.

Let M be a left $\mathcal{D}$ module of finite type supported by $\{0\}$; we define its cohomology:

$$H^n(M) = \frac{M}{\sum \partial_i M} \quad .$$

If N is another left $\mathcal{D}$ module of finite type supported by $\{0\}$ and $\phi: M \to N$ is a morphism of $\mathcal{D}$ modules, we associate to it

$$H^n(\phi): H^n(M) \to H^n(N), \quad (\dot{m}) \mapsto \phi(\dot{m}).$$

Proposition 2.2 *[M 2] Let M be a left $\mathcal{D}$ module of finite type supported by* $\{0\}$:

1.
$$M = 0 \Leftrightarrow H^n(M) = 0 \; ;$$

2. $H^n(M)$ *is a finite dimensional vector space over* $\mathbf{C}$.

If $0 \to M' \xrightarrow{\phi} M \xrightarrow{\psi} M" \to 0$ is an exact sequence on left $\mathcal{D}$ modules of finite type supported by $\{0\}$, the sequence

$$0 \to H^n(M') \xrightarrow{H^n(\phi)} H^n(M) \xrightarrow{H^n(\psi)} H^n(M") \to 0$$

is exact.

2.3.   A $\mathcal{D}$ module whose characteristic variety is reduced to $T_Y^* \mathbf{C}^n$ (Y smooth)

Let Y be a smooth analytic variety in a neighborhood of 0. We choose a set of coordinates $(x_1, \ldots, x_n)$ such that Y is locally given by the equations $x_1 = \ldots = x_k = 0$. Then,

$$T_Y^* \mathbf{C}^n = \{(0, \ldots, 0, x_{k+1}, \ldots, x_n); (\xi_1, \ldots, \xi_k, 0 \ldots, 0) \in U \times \mathbf{C}^n\}$$

is a subvariety of $T^* U = U \times \mathbf{C}^n$, where U is some neighborhood of 0.

Notation 1
$$B_{Y|\mathbf{C}^n} = \frac{\mathcal{D}}{\mathcal{D}(x_1, \ldots, x_k, \partial_{k+1}, \ldots, \partial_n)} \; .$$

An easy computation shows that the characteristic variety of $B_{Y|\mathbf{C}^n}$ is $T_Y^* \mathbf{C}^n$.

Proposition 2.3 *Let M be a finite left $\mathcal{D}$ module, the following assertions are equivalent*

- $\text{char}(M) = T_Y^* \mathbf{C}^n$;

- M is isomorphic as $\mathcal{D}$ module to a finite direct sum of copies of $B_{Y|\mathbf{C}^n}$.

Proof :   In the previous subsection, we have proved the case when $Y = 0$. The general case is more difficult. It is a consequence of a classical theorem of Cauchy with parameters. We refer to [Bjo] [G.M.].

Remark: $\mathcal{O}$ is a $\mathcal{D}$ module isomorphic to $B_{\mathbf{C}^n|\mathbf{C}^n}$. Indeed we have the following isomorphism of $\mathcal{D}$ modules:

$$ B_{\mathbf{C}^n|\mathbf{C}^n} = \frac{\mathcal{D}}{\mathcal{D}(\partial_1,\ldots,\partial_n)} \to \mathcal{O} , \quad \dot{P} \mapsto P(f) . $$

We also remark that the $\mathcal{O}$ modules with an integral connection are exactly the left $\mathcal{D}$ modules of finite type whose characteristic variety is $T^*_{\mathbf{C}^n}\mathbf{C}^n$. The proposition is equivalent to the existence of horizontal sections in an integrable connection.

Lemma 2.3 *We have the following isomorphism:*

$$ \mathbf{C} \simeq Hom_{\mathcal{D}}(B_{Y|\mathbf{C}^n}, B_{Y|\mathbf{C}^n}), $$

$$ \lambda \mapsto \text{multiplication by } \lambda . $$

Proof :   The proof is similar to the proof of Lemma 2.2.

Proposition 2.4 *Let M be a left $\mathcal{D}$ holonomic module and ϕ an endomorphism of M as a left $\mathcal{D}$ module. Then ϕ has a minimal polynomial*

$$ \exists c(s) \in \mathbf{C}[s] - \{0\} \quad such \quad that \quad c(\phi)(M) = 0 . $$

Proof :   We only give a sketch of proof. First we recall some facts from the paper by Malgrange. The variety $\mathrm{char}(M)$ has a finite number of irreducible components. Let Λ_i be one of them, and π be the projection of $T^*\mathbf{C}^n$ on $\mathbf{C}^n$. Λ_i is an analytic germ of $T^*\mathbf{C}^n$. $Y_i = \pi(\Lambda_i)$ is a germ of analytic space of $\mathbf{C}^n$. Let Y_i° be the smooth part of Y_i.
We denote by $T^*_{Y_i}\mathbf{C}^n$ the closure of $N^\circ_{Y_i}$ in $T^*\mathbf{C}^n$ and $N^\circ_{Y_i}$ is the set $\{(x,\xi) \in T^*\mathbf{C}^n\}$ such that $x \in Y_i^\circ$ and ξ is zero of the tangent to Y_i in x.
M is holonomic and the involuteness of $\mathrm{char}(M)$ implies that $\Lambda_i = T^*_{Y_i}\mathbf{C}^n$. Therefore we can consider the decomposition into irreducible components of $\mathrm{char}(M)$ writing

$$ \mathrm{char}(M) = \cup_{i \in I} T^*_{Y_i}\mathbf{C}^n , $$

where Y_i are irreducible analytic germs of $\mathbf{C}^n$.

M is holonomic and we have shown in the previous section the existence of a left ideal I of $\mathcal{D}$ such that $M \simeq \mathcal{D}/I$. Let $(P_1, \ldots, P_r)$ be a system of generators of I. The P_i are defined on a neigborhood U of 0. These P_i generate also an ideal I_{x_0} of $\mathcal{D}_{x_0}$, where $\mathcal{D}_{x_0}$ denotes the differential operators whose coefficients are analytic in a neigborhood of x_0. Let x_0 be a generic point on the smooth part of Y_j (Y_j is chosen with maximum dimension). We can prove that the $\mathcal{D}_{x_0}$ module $\mathcal{D}_{x_0}/I_{x_0}$ has only $T_{Y_i}^* \mathbf{C}^n$ as characteristic variety in a neigborhood of x_0. By the previous lemma, there exists a polynomial $e(s)$ such that $e(\phi)(M_{x_0}) = 0$. We consider $e(\phi)(M)$. It is a $\mathcal{D}$ submodule of M, so it is an holonomic $\mathcal{D}$ module and then its characteristic variety is included in $\text{char}(M)$. But we can prove that $T_{Y_j}^* \mathbf{C}^n$ is not a component of $\text{char}(e(\phi)(M))$. Therefore we obtain that for a set $J \subset I$ and $J \neq I$,

$$\text{char}(e(\phi)(M)) = \cup_{i \in J} T_{Y_i}^* \mathbf{C}^n \quad .$$

We repeat this process with $e(\phi)(M)$, ... and we are done when we find an empty characteristic variety.

Remark A precise proof needs sheaf theory or an algebraic theory of localization.

Now we associate to a left $\mathcal{D}$ module M the $\mathbf{C}$ vector space:

$$H^n(M) = \frac{M}{\sum \partial_i M} \quad .$$

If

$$0 \to M' \to M \to M" \to 0$$

is an exact sequence of left $\mathcal{D}$ modules, we have an exact sequence:

$$H^n(M') \to H^n(M) \to H^n(M") \to 0 \quad ,$$

where the morphisms are natural. For example with the theorem of structure and a little homological algebra, we can prove (for another proof see [M 2]).

Proposition 2.5 *If*
$$0 \to M' \to M \to M" \to 0$$

is an exact sequence of left $\mathcal{D}$ modules and we assume that $M"$ is supported by 0, we have the exact sequence

$$0 \to H^n(M') \to H^n(M) \to H^n(M") \to 0 \quad .$$

2.4.   Introduction to Bernstein polynomials

Let s be a new variable and $\mathcal{D}[s]$ denote the ring $C[s] \otimes_C \mathcal{D}$. Let $f \in C\{x_1, \ldots, x_n\} = \mathcal{O}$; we consider the ring $\mathcal{O}[1/f, s]$ whose elements can be written in the following form:

$$\sum_{i,j} a_{i,j} s^i f^j \quad \text{such that} \quad a_{i,j} \in \mathcal{O}, \quad i \in \mathbf{N}, \quad j \in -\mathbf{N}.$$

Then we consider the $\mathcal{O}[1/f, s]$ free module $\mathcal{O}[1/f, s]f^s$. It has a natural structure of $\mathcal{D}$ module given by

$$\partial_i(af^s) = \partial_i(a)f^s + s\frac{a}{f}\partial_i(f)f^s.$$

Theorem 2.4 *[Ber][K 2] The ideal of $C[s]$*

$$I = \{e(s) \in C[s] \quad ; \quad e(s)f^s \in \mathcal{D}[s]f^{s+1}\}$$

is not reduced to $\{0\}$.

Definition 2.2 *The Bernstein polynomial b_f of f is the unitary generator of I.*

Example

$$(s+1)(s+\frac{n}{2})(\sum x_i^2)^s = \frac{1}{4}(\sum \partial_i^2)(\sum x_i^2)^{s+1}.$$

Remarks

1. It is easy to prove that

$$f(0) \neq 0 \iff b_f = 1.$$

2. If the hypersurface $f = 0$ is smooth and $f(0) = 0$, there exists i such that $\partial_i(f)(0) \neq 0$. Hence, we have the functional equation

$$\frac{1}{\partial_i(f)}\partial_i f^{s+1} = (s+1)f^s,$$

and $b_f(s) = s+1$. The converse is true, see [B.M. 2].

3. $f(0) = 0$ implies that b_f is divisible by $s+1$ (take $s = -1$ in the functional equation associated to b_f). Then we set

$$b_f(s)/(s+1) = \tilde{b}_f(s).$$

Now, we give a sketch of Kashiwara's proof of the existence of b_f .

- In the case where f is in the jacobian ideal $(\partial_1(f),\ldots,\partial_n(f))\mathcal{O}$, we write $f = \sum_{i=1}^{n} a_i\partial_i(f)$ and we denote by χ the differential operator $\chi = \sum_{i=1}^{n} a_i\partial_i$. We have

$$\chi(f) = f \quad , \quad \mathcal{D}[s]f^s = \mathcal{D}f^s \quad , \quad (\partial_i(f)\partial_j - \partial_j(f)\partial_i)f^s = 0 \quad .$$

Then in a point near 0, the dimension of the characteristic variety of the $\mathcal{D}_{x_0}$ module generated by f^s is smaller than $n+1$.
With algebra and sheaf tools, one can prove that

$$\forall u \in \mathcal{D}f^s \quad , \quad \exists k \in \mathbf{N} \quad ; \quad f^k u \in L \quad ,$$

with

$$L = \{u \in \mathcal{D}f^s \quad ; \quad \dim \operatorname{char}(\mathcal{D}u) \leq n+1\} \quad .$$

If we apply this to $u = f^s$ we find k such that

$$\dim \operatorname{char}(\mathcal{D}f^{s+k}) \leq n+1 \quad .$$

But viewed as a $\mathcal{D}$ module, $\mathcal{D}f^{s+k}$ is isomorphic to $\mathcal{D}f^s$. Therefore $\dim(\operatorname{char}(\mathcal{D}f^s)) \leq n+1$.
Consider the exact sequence

$$0 \to \mathcal{D}f^{s+1} \to \mathcal{D}f^s \to \mathcal{D}f^s/\mathcal{D}f^{s+1} \to \quad 0.$$

Since $\mathcal{D}f^s$ is isomorphic to $\mathcal{D}f^{s+1}$, by the multiplicity argument explained in the previous chapter, we obtain that $\mathcal{D}f^s/\mathcal{D}f^{s+1}$ is a holonomic $\mathcal{D}$ module. The multiplication by s in $\mathcal{D}f^s/\mathcal{D}f^{s+1}$ is an endomorphism of left $\mathcal{D}$ modules, by Proposition 2.4, it has a minimal polynomial. This minimal polynomial is b_f.

- When f is not in the jacobian ideal, we use a trick: $g(x,u) = e^u f(x)$, then $\partial/\partial u(g(x,u)) = g$. Hence b_g exists. The existence of b_f can be derived from that of b_g.

2.5. Computation of b_f when f has an isolated singularity

We assume $f(0) = 0$ and f has an isolated singularity at 0; i.e.

$$\partial_1(f)(x) = \ldots = \partial_n(f)(x) = 0 \Rightarrow x = (0,\ldots,0) \quad .$$

We denote by $J(f)$ the jacobian ideal and by μ the Milnor number of f:

$$J(f) = (\partial_1(f), \ldots, \partial_n(f)) \quad , \quad \mu = \dim_{\mathbf{C}}(\mathcal{O}/J(f)) \quad .$$

We recall that if M is a left $\mathcal{D}$ module, then $H^n(M)$ is the $\mathbf{C}$ vector space $M/\partial_i M$. From [M 2] and the structure theorem of regular $\mathcal{D}$ modules in one variable 1.4., we have the proposition

Proposition 2.6 *(B. Malgrange) The canonical injection*

$$\mathcal{D}f^s + \ldots + \mathcal{D}s^{\mu-1}f^s \hookrightarrow \mathcal{D}[s]f^s$$

induces an isomorphism of $\mathbf{C}$ *vector spaces:*

$$H^n(\mathcal{D}f^s + \ldots + \mathcal{D}s^{\mu-1}f^s) \hookrightarrow H^n(\mathcal{D}[s]f^s) \quad .$$

From this result, we show how to compute b_f.

Lemma 2.5 $\mathcal{D}[s]f^s/\mathcal{D}f^s + \ldots + \mathcal{D}s^{\mu-1}f^s$ *is supported by* 0.

Proof :   Let $\mathcal{M}$ be the maximal ideal of $\mathcal{O}$. By the Nullstellensatz :

$$\exists k \in \mathbf{N} \quad ; \quad \mathcal{M}^k \in J(f) \quad .$$

The lemma follows from the equality

$$(s+1)\partial_i(f)f^s = \partial_i f^{s+1} \quad .$$

From the previous proposition, it results from direct computation that

$$H^n(\mathcal{D}[s]f^s/\mathcal{D}f^s + \ldots + \mathcal{D}s^{\mu-1}f^s) = 0 \quad .$$

If we use the structure of $\mathcal{D}$ modules supported by a point (subsection 2.2.), we obtain

Proposition 2.7 $\mathcal{D}[s]f^s$ *is a* $\mathcal{D}$ *module of finite type, more precisely*

$$\mathcal{D}[s]f^s = \mathcal{D}f^s + \ldots + \mathcal{D}s^{\mu-1}f^s \quad .$$

By definition $\tilde{b}_f(s) = b_f(s)/(s+1)$ is the minimal polynomial of the action of s on $(s+1)(\mathcal{D}[s]f^s/\mathcal{D}[s]f^{s+1})$.

Lemma 2.6

$$(s+1)(\mathcal{D}[s]f^s/\mathcal{D}[s]f^{s+1}) \simeq \mathcal{D}[s]f^s/\mathcal{D}[s](f,J(f))f^s \ ,$$

$$(s+1)Pf^s \mapsto Pf^s \ \ .$$

Proof : The only difficulty is to prove that this application is well defined. Let $P \in \mathcal{D}[s]$ such that $(s+1)Pf^s = Q(s)f^{s+1}$ where $Q(s) \in \mathcal{D}[s]$. We write

$$Q(s) = (s+1)T(s) + R \quad \text{where} \quad T(s) \in \mathcal{D}[s] \quad , \quad R \in \mathcal{D} \ \ .$$

Take $s = -1$, we find $R(1) = 0$. Then we can write $R = \sum R_i \partial_i$. And we obtain

$$Pf^s = (Q(s)f + \sum R_i \partial_i(f))f^s \ \ .$$

Hence, the application is well defined.

By the previous lemma and the structure theorem, we have the following proposition:

Proposition 2.8 *(B. Malgrange)* $\tilde{b}_f(s)$ *is the minimal polynomial of the action of* s *on*

$$H^n(\mathcal{D}[s]f^s/\mathcal{D}[s](f,J(f))f^s) \simeq H^n(\mathcal{D}[s]f^s)/H^n(\mathcal{D}[s](f,J(f))f^s) \ \ .$$

We now take μ elements **C** independent of $\mathcal{O}$ ($e_1, \ldots, e_\mu$) which induce a basis of $\mathcal{O}/J(f)$. We let

$$E = \sum \mathbf{C} e_i \subset \mathcal{O} \ \ .$$

Then we have

$$(*) \quad \forall u \in \mathcal{O} \quad u = e + \sum u_i \partial_i(f) \quad \text{with} \quad e \in E \ , \quad u_i \in \mathcal{O} \ \ .$$

We let

$$\xi_i = s(s-1)\ldots(s-i+1)f^{s-i} \ ,$$

$$\tilde{D}E = \{\sum \partial^\alpha e(\alpha) \ ; \ e(\alpha) \in E\} \subset \mathcal{D} \ \ .$$

Proposition 2.9 *We have the following decomposition into a direct sum:*

$$\sum \mathcal{D}\xi_i = \mathcal{D}J(f)f^s \oplus \tilde{D}E\xi_0 \oplus \ldots \oplus \tilde{D}E\xi_i \oplus \ldots \ \ .$$

Proof : Iterate (∗) and use the following lemma (see [B.G.M.M.] for a precise proof):

Lemma 2.7 *If P is a differential operator independent of s*

$$Pf^s = 0 \Leftrightarrow P \in \mathcal{D}(\partial_i(f)\partial_j - \partial_j(f)\partial_i) \quad .$$

Proof : It is clear that $(\partial_i(f)\partial_j - \partial_j(f)\partial_i)f^s = 0$. The converse results from the properties of the sequence $(\partial_1(f), \ldots, \partial_n(f))$ which is a regular sequence ([Mat]). The key point is that if $\phi(\lambda_1, \ldots, \lambda_n) \in \mathcal{O}[\lambda_1, \ldots \lambda_n]$ is a polynomial homogeneous in λ with

$$\phi(\partial_1(f), \ldots, \partial_n(f)) = 0$$

then $\phi(\lambda_1, \ldots, \lambda_n)$ is divisible by $\partial_i(f)\lambda_j - \partial_j(f)\lambda_i$.
From Lemma 2.7, we conclude:

Lemma 2.8 $\mathcal{D}[s]f^s / \mathcal{D}[s](f, J(f))f^s$ *and* $\sum \mathcal{D}\xi_i / \mathcal{D}[s]f^s$ *are supported at* 0.

From Lemma 2.7, we deduce the injections

$$H^n(\mathcal{D}[s](f, J(f))f^s) \hookrightarrow H^n(\mathcal{D}[s]f^s) \hookrightarrow H^n(\sum \mathcal{D}\xi_i) \quad .$$

From Proposition 2.9, we deduce

$$H^n(\sum \mathcal{D}\xi_i) = H^n(\mathcal{D}J(f)f^s) \oplus E\xi_0 \oplus \ldots \oplus E\xi_i \oplus \ldots \quad .$$

We denote by c the canonical projection:

$$H^n(\mathcal{D}J(f)f^s) \oplus E\xi_0 \oplus \ldots \oplus E\xi_i \oplus \ldots \rightarrow E\xi_0 \oplus \ldots \oplus E\xi_i \oplus \ldots$$

By Proposition 2.8, we obtain $\tilde{b}_f$ as the minimal polynomial of an action of s on

$$\frac{c(H^n(\mathcal{D}[s]f^s))}{c(H^n(\mathcal{D}[s](f, J(f))f^s))} \quad .$$

Remark: The computation of this polynomial is algorithmic.
Let k be such that $\mathcal{M}^k \subset J(f)$, then $c(\mathcal{M}^{k+(j-1)(k+1)}s^j f^s) = 0$. Denote by $\mathcal{O}_{\leq j}$ the subset of $\mathcal{O}$ consisting of the polynomials of degree $\leq j$. We have

$$
\begin{aligned}
c(H^n(\mathcal{D}[s]f^s)) &= & c(H^n(\mathcal{D}f^s + \ldots + \mathcal{D}s^{\mu-1}f^s)) \\
&= & c(H^n(\mathcal{O}f^s + \ldots + \mathcal{O}s^{\mu-1}f^s)) \\
&= & c(H^n(\mathcal{O}_{\leq k}f^s + \ldots + \mathcal{O}_{\leq k+(\mu-1)(k+1)}s^{\mu-1}f^s))
\end{aligned}
$$

We have a similar result for $c(H^n(\mathcal{D}[s](f, J(f))f^s))$. These vector spaces are of finite dimension. The determination of the subvectorspaces

$$c(H^n(\mathcal{D}[s](f, J(f))f^s)) \quad \text{and} \quad c(H^n(\mathcal{D}[s]f^s))$$

of $\sum_{i=1}^{\mu} E\xi_i$ can be achieved after a finite number of computations. With standard basis technique, all computations are algorithmic. Thus, we have produced an algorithm to compute the Bernstein polynomial for an isolated singularity.

Proposition 2.10 *Let* $\tilde{c}(s) = \tilde{b}_f(s) \ldots \tilde{b}_f(s + \mu - 1)$ *and* $c(s) = (s + 1)\tilde{c}(s)$

1. *We have the following precise functional equations:*

    $\exists A \in \mathcal{D}$ *(independent of* s*) such that* $c(s)f^s = Af^{s+1}$.

    and $degree(A) \leq deg(c(s)) = \mu(\mu - 1) + 1$,

2. *We have the following precise operator in the annihilator of* f^s*:*

    $$(\tilde{c}(s) - B)f^s = 0$$

    with $B \in \mathcal{D}$ *independent of* s *and degree* $(B) \leq deg(\tilde{c}(s)) = \mu(\mu - 1)$.

Proof :   First, it is clear that

$$(*)\quad b_f(s) \ldots b_f(s + \mu - 1)f^s = P(s)f^{s+\mu}, \quad P(s) \in \mathcal{D}[s] .$$

By Proposition 2.7, there are differential operators $A_1, \ldots, A_\mu$ in $\mathcal{D}$ such that

$$(s^\mu + A_1 s^{\mu-1} + \ldots + A_\mu)f^s = 0 ,$$

but the degree of A_i is not determined. By dividing $P(s)$ by the operator $s^\mu + A_1 s^{\mu-1} + \ldots + A_\mu$ we can assume that the degree in s of $P(s)$ is $\leq \mu - 1$. Then, we write $P(s) = (s + \mu)Q(s) + R$, where $Q(s) \in \mathcal{D}[s]$ with degree in s being $\leq \mu - 2$ and where $R \in \mathcal{D}$. Putting $s = -\mu$ in $(*)$, we obtain $0 = R(1)$, then $R = \sum R_i \partial_i$. We write $(*)$:

$$b_f(s) \ldots \tilde{b}_f(s + \mu - 1)f^s = (Q(s)f + \sum R_i \partial_i(f))f^{s+\mu-1} .$$

After μ steps, we obtain an operator $L \in \mathcal{D}$ independent of s, such that $c(s)f^s = Lf^{s+1}$. Assume that the degree of L is $\geq$ the degree in s of $c(s)$. We can write

$$L = \sum_{k=0}^{k=deg(L)} \sum_{|\alpha|=k} p_\alpha \partial^\alpha \quad \text{where} \quad p_\alpha \in \mathcal{O} .$$

We compute $c(s)f^s = Lf^{s+1}$; the terms of highest degree in s give

$$0 = \sum_{|\alpha|=deg(L)} p_\alpha (\partial_1(f))^{\alpha_1} \ldots (\partial_n(f))^{\alpha_n}$$

But with an algebraic argument similar to the one used in the proof of Lemma 2.7

$$\sum_{|\alpha|=deg(L)} p_\alpha \xi^\alpha = \sum \Lambda_{i,j}(\partial_i(f))\xi_j - \partial_j(f))\xi_i) \quad \text{with} \quad \Lambda_{i,j} \in \mathcal{O}[\xi] \quad .$$

Now, we pull back $\Lambda_{i,j}$ and we obtain $P_{i,j}$ with principal symbol equal to $\Lambda_{i,j}$. The degree of the operator

$$L' = L - \sum P_{i,j}(\partial_i(f)\partial_j - \partial_j(f)\partial_i)$$

satisfies $deg(L') \leq deg(L)$ and we still have $c(s)f^s = L'f^{s+1}$. We repeat this process if the degree of L' is $\leq$ the degree in s of $c(s)$ ... and we obtain the first part of the proposition.

In the equation obtained $c(s)f^s = Af^{s+1}$, we let $s = -1$ and we prove as previously that we can write $R = \sum A_i \partial_i$ and then $B = \sum A_i \partial_i(f)$ satisfies the required conditions.

2.6.   Basics on regularity in several variables

Let H be the hypersurface of $\mathbf{C}^n$ with equation $x_n = 0$. We define the V filtration indexed by $\mathbf{Z}$ of the ring $\mathcal{D}$ of differential operators:

$$V_j = \{P \in \mathcal{D} \quad ; \quad \forall k \in \mathbf{N} \quad PJ^k \subset J^{k-j}\}$$

with $J = x_n \mathcal{O}$.

- $\partial_n \in V_1\mathcal{D}$, $t \in V_{-1}\mathcal{D}$

- for $i \neq n$: x_i and $\partial_i \in V_0\mathcal{D}$

- $\sum a_{\alpha,j}(x)(x_n\partial_n)^j (\partial_1)^{\alpha_1} \ldots (\partial_{n-1})^{\alpha_{n-1}} \in V_0\mathcal{D}$

Following Y. Laurent we define:

Definition 2.3 *M a left $\mathcal{D}$ module is said to be specializable relative to H if $\forall u \in M$, $\exists b(s) \in \mathbf{C}[s] - \{0\}$ and $\exists P \in V_{-1}(\mathcal{D})$ such that $(b(x_n\partial_n) + P)u = 0$. Let r be a rational number; M is said to be r-specializable if $\forall u \in M$, $\exists b(s) \in \mathbf{C}[s]$ and an equation*

$$b(x_n\partial_n) + \sum t^k P_k(x, \partial_1, \ldots, \partial_{n-1}, x_n\partial_n)u = 0$$

with $deg_{\partial_1, \ldots \partial_n}(P_k) \leq deg_s(b) + (r-1)k$.

Theorem 2.9 *[K.K.] A holonomic left $\mathcal{D}$ module is 1-specializable along all hypersurfaces.*

For $r = 1$, 1-specializable generalized the notions of regularity. This condition is also equivalent to the existence of an isomorphism between any convergent solutions and any formal solutions. We refer to [K.K. 2][Meb].

The notion of r-specialization generalize the notion of the irregularity of an operator in one variable. A difficult task is to find a relation between this definition and basic properties of some "solutions" of M. A result of that kind is proved by Y. Laurent in [Lau 2].

References

[Ber]   I.N. Bernstein, *The analytic continuation of generalized functions with respect to a parameter*, Fonct. Anal. and Applic., 6 (1972).

[Bjo]   J.E. Bjork, *Rings of differential operators*, North Holland, 1979.

[Bou]   L. Boutet de Monvel, *D-modules holonomes réguliers en une variable*. Séminaire E.N.S. 1979-82, Progress in Math., 2, Acad. Press, 1987.

[B.G.M.] J.Briançon, F.Geandier, Ph. Maisonobe, *Déformation d'une singularité isolée d'hypersurface et polynôme de Bernstein*, Bulletin de la Soc.Math.France , 120, (1992), 15-49.

[B.G.M.M.] J. Briançon, M. Granger , Ph. Maisonobe , M. Miniconi, *algorithme de calcul du polynôme de Bernstein: cas non dégénéré*, Annales de l'Institut Fourier, 39, Fascicule 3, (1989), 553 - 610.

[B.M.]   J. Briançon and Ph. Maisonobe, *Idéaux de germes d'opérateurs à une variable.*, L'Enseignement Mathématique, 30, (1984), 7-38.

[B.M. 2] J. Briançon and Ph. Maisonobe, *Autour d'une conjecture sur les $\mathcal{D}$ modules holonomes réguliers, cohérents relativement à une projection*, Prépublication 290 de l'Université de Nice (1990).

[C]   F. Castro, Thèse de 3ème cycle, Université de Paris 7 (1984).

[G]   A. Galligo, *Some algorithmical questions on ideals of differential operators*, Lect. Notes Comput. Sci., Springer, 204, (1985), pp. 1452-1460.

[Gr]   D.Y. Grigorev, *Complexity of solving systems of linear equations over the rings of differential operators, an Effective Method in Algebraic Geometry,*, Progress in Mathematics, 94, (1991).

[G.M.]   M. Granger and Ph. Maisonobe, *A basic course on differential modules, Eléments sur les systèmes différentiels*, Cours du C.I.M.P.A., Hermann, to appear.

[K]   M. Kashiwara, *On the maximally overdetermined systems of linear differential equations 1*, Pub. R.I.M.S. Kyoto Uni.(1975), 563-569.

[K 1]   M. Kashiwara, *Systems of microdifferential equations*, Progress in Math., 34, Birkhauser.

[K 2]   M. Kashiwara, *B functions and holonomic sytems*, Inventiones Mathematicae, 38 (1976).

[K 3]   M. Kashiwara, *On the holonomic systems of linear differential equations II*, Inventiones Mathematicae, 49 (1978).

[K.K.]   M. Kashiwara, T. Kawai, *Second Microlocalisation and asymptotic expansions*, Lecture Notes in Physics, 126, Springer , (1980), 21-76.

[K.K. 2] M. Kashiwara, T. Kawai, *On the holonomic systems of microdifferential equations 3, Systems with regular singularities*, Publ. R.I.M.S., Kyoto Univ. 17 (1981), 813-979.

[Lau 1]   Y.Laurent, *Polygone de Newton et b. Fonctions pour les modules microdifférentiels*, Scient. Ec. Norm. Sup. $4^{\text{ème}}$ série, 20, (1987), 391-441.

[Lau 2]   Y.Laurent, *Vanishing cycles of $\mathcal{D}$ modules*, Prépublication de l'Institut Fourier, Grenoble (1991).

[L.M]   M. Lejeune and B. Malgrange, *Séminaire sur les opérateurs différentiels*, preprint (Grenoble 1976).

[M 1]   B. Malgrange, *Sur les points singuliers des équations différentielles*. L'Enseignement Mathématique, 20 (1974), 146-176.

[M 2]   B. Malgrange, *Le polynôme de Bernstein d'une singularité isolée*, Lect. Notes Math., 459, Springer, (1975), 98-119.

[M 3]   B. Malgrange, *Polynôme de Bernstein-Sato et cohomologie évanescente*, Astérisque 101-102, 233-267 (1983).

[M 4]   B. Malgrange, *Equations Différentielles à Coefficients Polynomiaux*, Progress in Math., 96, (1991).

[Mai]   Ph. Maisonobe, *Germes de D-modules à une variable et leurs solutions. Introduction à la théorie algébrique des systèmes différentiels*, 34, Hermann, (1989).

[Mat]   H. Matsumura, *Commutative ring theory*, Cambridge University Press *(1986)*.

[Meb]   Z. Mebkhout, *Le formalisme des 6 opérations de Grothendieck pour les $\mathcal{D}_X$ modules cohérents*, 35, Hermann(1989).

[M.N.]   Z.Mebkhout et L. Narvaez, *Le Théorème de constructibilité de Kashiwara, Cours du C.I.M.P.A.*, Hermann, (to appear) .

[P]   F. Pham, *Singularités des systèmes différentiels de Gauss Manin*, Progress in Math., 2, Birkhauser(1979).

[R]   J.P. Ramis, *Théorèmes d'indice Gevrey pour les équations différentielles ordinaires*, Mem. Amer. Math. Soc., 48 (1984) .

[S]   C. Sabbah, *Introduction to algebraic theory of linear systems of differential equations*, Eléments sur les systèmes différentiels, Cours du C.I.M.P.A., Hermann, to appear.

[W]   W. Wasow, *Asymptotic expansions for ordinary differential equations.*, Interscience, Publ. (1965).

Chapter 2

Theoretical aspects
in dynamical systems

E. Delabaere

Introduction to the Ecalle theory

K.R. Meyer

Perturbation analysis of nonlinear systems

Introduction to the Ecalle theory

Eric Delabaere [*]

Contents

1. **Introduction to resurgence theory**   61

   1.1. Introduction 61

   1.2. Simple resurgent functions 65

      1.2.1. The algebra of Gevrey order 1 power series 65

      1.2.2. Endlessly continuable functions 66

      1.2.3. Simple resurgent functions 69

      1.2.4. Elementary resurgent symbols 70

   1.3. Summations and connection automorphism 71

      1.3.1. Lateral Borel summations 71

      1.3.2. Connection automorphism and resurgent symbols . . . 72

   1.4. Alien derivatives 75

      1.4.1. Alien differentiations of resurgent symbols 75

      1.4.2. Commutation with the natural differentiation 76

      1.4.3. Resurgence equations 76

2. **Equational and quantum resurgence**   79

   2.1. First example 79

      2.1.1. Equational resurgence 79

      2.1.2. Quantum resurgence 85

[*]Member of CNRS. Department of Maths. U.A. CNRS N° 168. University of Nice, France.

2.2. Second example 88

    2.2.1. Equational resurgence 90

    2.2.2. Quantum resurgence 95

1.   Introduction to resurgence theory

This section is a very brief introduction to *resurgence*. Forgetting the purely technical difficulties, our aim is to present the noteworthy simple basic ideas of the Ecalle theory. In this way, we shall restrict ourselves to the quite simple algebra of *simple resurgent functions*, which gives a very pleasant context for beginning the theory.

The framework is the following. We begin by defining a subalgebra of the multiplicative algebra of formal power series $C[[x^{-1}]]$, furnishing through the Borel transformation a convolutive subalgebra of analytic germs at the origin. On the other hand, in order to sum by a Laplace transformation, the analytic continuation of these germs must have only "few" singularities, a notion which has to be stable under the convolution product. After having defined the algebra of simple resurgent functions, we get naturally the notion of *resurgent symbols* by a comparison of the different summations, in other words by an analysis of the *Stokes phenomena*. These can be described either with the help of an automorphism of algebra or with new differentiations, the *alien differentiations*.

A bibliography will allow the reader to go further into the theory. At first, we naturally send the reader to the whole work of Ecalle himself. We have followed here more or less the clear presentation of the reference [CNP] where one can find all the basic tools with complete proofs and some applications.

1.1.   Introduction

The Ecalle theory is based on the Borel process for the summation of divergent power series. Given the equality,

$$\text{For } n \in \mathbf{N}, \quad x^{-n-1} = \int_0^{+\infty} e^{-x\xi} \frac{\xi^n}{n!}\, d\xi,$$

this can be described by the diagram overleaf:

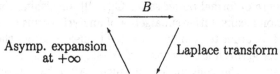

Formal power series $\varphi \in x^{-1}\mathbf{C}[[x^{-1}]]$

$$\varphi = \sum_{n\geq 0} \frac{a_n}{x^{n+1}}$$

Convergent expansion $\varphi \in \mathbf{C}\{\xi\}$

$$\varphi = \sum_{n\geq 0} \frac{\xi^n}{n!}$$

whose sum has an analytic continuation in a neighbourhood of $\mathbf{R}^+$ and a rate of growth less than an exponential at $+\infty$

Asymp. expansion at $+\infty$

Laplace transform

$$s\,\varphi(x) = \int_0^{+\infty} e^{-x\xi}\varphi(\xi)\,d\xi$$

analytic function in the half-plane $\Re e(x) \geq 0$.

The operator B is called the *Borel transformation* and $s\,\varphi$ is the *Borel sum* of the formal power series φ.

This Borel process brings up some natural questions, for instance:

i) How can we describe the analytic continuation of $s\,\varphi$ outside the half-plane $\Re e(x) \geq 0$?

ii) How can we define the sum of a formal power series whose Borel transform has singularities along the positive real axis $\mathbf{R}^+$?

The example of the Euler equation is in this matter particularly simple and instructive. Up to a change of variable, this differential equation is

$$\frac{\partial}{\partial x}\psi - \psi = \frac{-1}{x}.$$

Forget that this equation can be integrated and search for a solution as an entire expansion at infinity, without constant term. We easily get by identification the formal power series

$$\varphi = \sum_{n\geq 0} \frac{(-1)^n n!}{x^{n+1}}$$

which diverges for all x. Nevertheless its Borel transform

$$B\varphi = \sum_{n\geq 0}(-1)^n \xi^n$$

is a convergent expansion with a sum equal to $\varphi(\xi) = \frac{1}{1+\xi}$. The Borel sum $s\varphi$ is an analytic function in the half-plane $\Re e(x) > 0$, and one can check by differentiation under the sign of integration that it is a solution of the Euler equation, the one which vanishes at infinity.

Let us consider now the problem of the analytic continuation of the Borel sum $s\varphi$. Denote by α a direction in the ξ plane (the Borel plane) near the real positive direction and denote by $s_\alpha \varphi$ the integral

$$s_\alpha \varphi(x) = \int_{0_\alpha} e^{-x\xi}\varphi(\xi)\, d\xi$$

where the path of integration 0_α starts from the origin and goes to infinity along the direction α. This integral is convergent in the half-plane P_α of the x such that $\Re e(x\xi) > 0$.

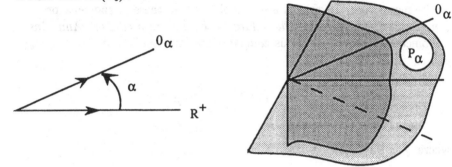

By Cauchy's theorem, the integrals $s_\alpha \varphi$ and $s\varphi$ coincide on the intersection of their convergence domains and thus are analytic continuations the one from the other. On the other hand, for x at infinity in the half-plane P_α, the asymptotic expansion of $s_\alpha \varphi$ is here again given by the power series φ.

We come now to the heart of the matter by considering the singular direction $\alpha = (\pi)$. We shall denote by 0_{α_+} (resp. 0_{α_-}) the endless path from the origin to infinity along the direction α and avoiding the singularity -1 towards $Arg\,\xi > Arg\,\alpha$ (resp. $Arg\,\xi < Arg\,\alpha$),

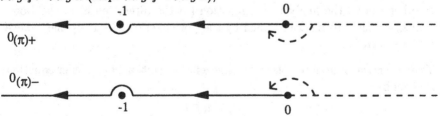

which allows us to define the integrals

$$s_{\alpha_\pm} \varphi(x) = \int_{0_{\alpha_\pm}} e^{-x\xi}\varphi(\xi)\, d\xi,$$

The integrals s_{α_+} and s_{α_-} are convergent in the same domain

$$P_\alpha = \{x/\Re e(x) < 0\}$$

and are both analytic continuations of $s\,\varphi$ around the origin, one in the positive way, in the negative way for the other. However they do not coincide because of the polar singularity of φ at -1 : the power series φ meets with a *Stokes phenomenon* in the direction α.

Note that this Stokes phenomenon implies that the function $s\,\varphi$ is an analytic *multivalued* function at the origin. It would be wrong to think that it is a general link: a formal power series may meet with some Stokes phenomena and its Borel sum be a uniform function. Nevertheless, the analysis of the monodromy is often a useful tool to make these phenomena precise (for instance, see A.Voros, *The return of the quartic oscillator*, Ann. Inst. H.Poincaré 29,3-1983-). Let us compare more precisely these two sums s_{α_+} and s_{α_-} : for $\Re e(x) < 0$,

$$(s_{\alpha_+} - s_{\alpha_-})\,\varphi(x) = \int_\gamma e^{-x\xi}\varphi(\xi)\,d\xi$$

where γ is the following path:

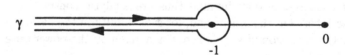

This gives

$$(s_{\alpha_+} - s_{\alpha_-})\,\varphi(x) = -2\pi\imath e^x.$$

In other words, the analytic continuation (in the direct sense) of the Borel sum $s_\alpha\,\varphi$ ceases to be determined by the power series φ, but by the *symbol* $\dot\psi = \varphi - 2\pi\imath e^x$.

To end this introduction, notice that the general solution of the Euler equation is given by

$$ue^x + s\varphi, \ \ u \in \mathbb{C},$$

and this allows us to think that it is all contained in essence in the formal power series φ through the analysis of the singularities of its Borel transform (see section 2).

1.2.   Simple resurgent functions

1.2.1.   The algebra of Gevrey order 1 power series

An expansion of *Gevrey order 1* [MR1] is a formal power series

$$\varphi = a_0 + a_1 x^{-1} + a_2 x^{-2} + a_3 x^{-3} + ... \in \mathbf{C}[[x^{-1}]]$$

such that $\left(\frac{|a_n|}{n!}\right)$ is less than a geometrical sequence. The set of these formal power series defines a unitary subalgebra of $\mathbf{C}[[x^{-1}]]$. Forgetting the constant term, a formal power series of Gevrey order 1 gives rise by the Borel transform to an analytic germ at the origin denoted by φ and called the *minor* of φ.

$$\varphi = a_0 + a_1 x^{-1} + a_2 x^{-2} + a_3 x^{-3} + ... \in \mathbf{C}[[x^{-1}]]$$

forget the constant term a_0

$$\downarrow$$

Borel transform

$$\varphi = a_1 + a_2 \frac{\xi}{1!} + a_3 \frac{\xi^2}{2!} + ... \in \mathbf{C}\{\xi\}$$

One often writes

$$\varphi = a_0 + {}^b\varphi,$$

where the flat symbol ^{b} is the *inverse of the Borel operator*. The constant term a_0 is called the *residual coefficient*. If the residual coefficient vanishes then we shall say that φ is a *small* power series (of Gevrey order 1). The usual product of two small power series becomes through the Borel transformation the convolution product of the minors:

$$^b\varphi.^b\psi \xrightarrow{B} \varphi * \psi,$$

$$\varphi * \psi(\xi) = \int_0^\xi \varphi(u)\psi(\xi - u)\, du,$$

where we integrate along the segment going from the origin to ξ.

Otherwise, the action on φ of the natural differentiation $\frac{\partial}{\partial x}$ becomes the action of the operator ∂ on the minor φ,

$$\frac{\partial}{\partial x} \xrightarrow{B} \partial$$

$$\partial \varphi(\xi) = -\xi \varphi(\xi).$$

Finally, in order to build a *unitary* convolutive subalgebra of the algebra $\mathcal{O}_0$ of analytic germs at the origin, we must define the Borel transform $B(1)$ of the unit of $\mathbf{C}[[x^{-1}]]$. Such a germ does not exist, so we add it abstractly by considering the convolutive algebra $\mathbf{C}\delta \oplus \mathcal{O}_0$ with $\delta * \varphi = \varphi$ and $\partial \delta = 0$.

Remark: The Borel transformation proceeds from the identity,

$$\text{For } n \in \mathbf{N}, \quad x^{-n-1} = \int_0^{+\infty} e^{-x\xi} \frac{\xi^n}{n!} d\xi = \int_C e^{-x\xi} \frac{\xi^n \log \xi}{2\pi i n!} d\xi,$$

where the path of integration C comes and goes back to $+\infty$ after having turned around the origin in the positive sense. As

$$1 = \int_C e^{-x\xi} \frac{1}{2\pi i \xi} d\xi,$$

the unit should be the "variation at the origin" of $\frac{1}{2\pi i \xi}$. The unity δ acts therefore as a Dirac function at the origin, hence the notation.

1.2.2.   Endlessly continuable functions

As we have seen in the introduction, the sum of a Gevrey 1 expansion will be defined provided that the analytic continuation of the minor on the complex plane meets only "few" singularities. The following definition explains what we mean by "few".

Definition 1.1 *A germ of analytic functions at the origin φ is said to be endlessly continuable on $\mathbf{C}$ if for all $L > 0$, there exists a finite set $\Omega_L(\varphi) \subset \mathbf{C}$ (the L-accessible singularities) such that φ has an analytic continuation along every path λ whose length is $< L$ and avoiding the set $\Omega_L(\varphi)$. In that case, its Riemann surface is said to be endless.*

Remark: This definition is not the most general one but it is enough for our framework. The minimal definition adopted by Ecalle is the notion of *functions everywhere continuable without cuts* [E1].

Thus the function $\varphi(\xi) = \frac{1}{1+\xi}$ given by the Euler equation at the very beginning is obviously endlessly continuable. But is this notion stable under convolution? Look at it in an example.

First example.   Let $\varphi(\xi) = \frac{1}{\xi - \alpha}$ and $\psi(\xi) = \frac{1}{\xi - \beta}$, where α and β belong to $C \backslash \{0\}$. The convolution product $\varphi * \psi$ is the germ of an analytic function at the origin defined by

$$\varphi * \psi(\xi) = \frac{1}{\xi - (\alpha + \beta)} \left(\int_0^\xi \frac{1}{\xi - \alpha} d\xi + \int_0^\xi \frac{1}{\xi - \beta} d\xi \right).$$

- If α and β are not on a half-line starting from the origin, then the germ $\varphi * \psi$ has an analytic continuation in the star domain

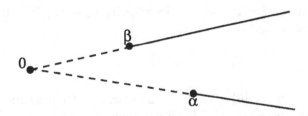

where α and β are logarithmic singularities. Moreover, there is a pole at $\alpha + \beta$ which does not exist on the first sheet, whose residue depends on the path of analytic continuation.

- If α and β are on the same half-line starting from the origin, then the germ $\varphi * \psi$ has an analytic continuation in the cut complex plane,

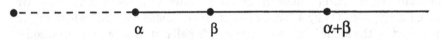

Here again there are logarithmic singularities at α and β and $\alpha + \beta$ is a pole whose residue depends on the path of analytic continuation.

Thus one sees that the convolution product gives birth to a new singularity. Nevertheless, even if the singularity may be near the origin by projection, it is in fact present "away" on the Riemann surface of the analytic continuations of the germ $\varphi * \psi$. This result extends to the general case in the following manner [CNP]:

Proposition 1.1 Let φ (resp. ψ) be a germ of analytic functions at the origin, endlessly continuable on C. For all $L > 0$, we denote by $\Omega_L(\varphi)$ (resp. $\Omega_L(\psi)$) the set of all L-accessible singularities of φ (resp. ψ). Then the germ $\varphi * \psi$ is endlessly continuable and the set $\Omega_L(\varphi * \psi)$ of its L-accessible singularities is included in the set

$$(\Omega(\varphi) + (\Omega(\psi))_L \overset{def}{=}$$

$$\{\omega = \omega_1 + \omega_2 \in \mathbf{C}, \ \omega_1 \in \Omega_{L_1}(\varphi) \ and \omega_2 \in \Omega_{L_2}(\psi) \ such \ that L_1 + L_2 \le L\}.$$

Note here that an endlessly Riemannian surface may have a *dense* set of singularities by projection on the complex plane! Thus this notion allows the major practical cases. Otherwise as we shall see in the following example, the preceding proposition often enables us to foresee the position of the singularities of the Borel transform of a formal power series, and in passing to prove the endless continuation.

Second example. Let us consider the following equation, formally conjugate to the Euler equation:

$$\frac{\partial}{\partial x}\psi - \lambda\psi = -\frac{1}{x} - \frac{1}{x^2}\psi^2 \tag{1}$$

where λ belongs to $\mathbf{C}\backslash\{0\}$. We begin by searching for a solution in the shape of a formal power series at infinity, without constant term. We get by identification,

$$\varphi(x, \lambda) = \lambda^{-1}x^{-1} - \lambda^{-2}x^{-2} + 2\lambda^{-3}x^{-3} + (-6\lambda^{-4} + \lambda^{-3})x^{-4}$$

$$+ (24\lambda^{-5} - 6\lambda^{-4})x^{-5} + \dots$$

In fact as often, the knowledge of this (small) formal power series is not useful for analysing its Borel transform and it is better to proceed otherwise. We solve the equation (1) by a "successive approximations method" which can be described by the following: we introduce a "small constant" ε and we consider the expansion

$$\varphi(x, \lambda, \varepsilon) = \varphi_{(0)}(x, \lambda) + \varepsilon\varphi_{(1)}(x, \lambda) + \varepsilon^2\varphi_{(2)}(x, \lambda) + \dots .$$

Then we formally solve the equation (2) by identification in ε :

$$\frac{\partial}{\partial x}\psi - \lambda\psi = -\frac{1}{x} - \frac{\varepsilon^2}{x^2}\psi^2; \tag{2}$$

and finally we write $\varepsilon = 1$. If we denote by $\mathcal{D}$ the operator $(\frac{\partial}{\partial x} - \lambda)$, we get for example for the first terms,

$$\mathcal{D}\varphi_{(0)} = -x^{-1} \ , \ \mathcal{D}\varphi_{(1)} = 0 \ , \ \mathcal{D}\varphi_{(2)} = -x^{-2}\varphi_{(0)}^2 \ ,$$

$$\mathcal{D}\varphi_{(3)} = -2x^{-2}\varphi_{(0)}\varphi_{(1)} \ , \ \mathcal{D}\varphi_{(4)} = -x^{-2}\varphi_{(1)}^2 - 2x^{-2}\varphi_{(0)}\varphi_{(2)},$$

$$\dots$$

Translating to the minors, we get

$$\varphi_{(0)} = (\xi + \lambda)^{-1} \, , \; \varphi_{(1)} = 0 \, , \; \varphi_{(2)} = (\xi + \lambda)^{-1}(\xi * \varphi_{(0)} * \varphi_{(0)}) \, ,$$

$$\varphi_{(3)} = 0 \, , \; \varphi_{(4)} = (\xi + \lambda)^{-1}(2\xi * \varphi_{(0)} * \varphi_{(2)}) \, ,$$

$$\cdots$$

The analytic germ $\varphi_{(0)}$ has $-\lambda$ as its unique singularity and we deduce from the previous proposition that $\varphi_{(2)}$ is an analytic function on the universal covering of $C\backslash\{-\lambda, -2\lambda\}$; therefore $\varphi_{(4)}$ is analytic on the universal covering of $C\backslash\{-\lambda, -2\lambda, -3\lambda\}$, ... and so on. Thus we see, admitting the convergence (see [CNP]), that the minor φ of the formal expansion φ defines a germ of an analytic function at the origin which extends to the universal covering of $C\backslash\{-\lambda, -2\lambda, -3\lambda, \ldots\}$. The formal power series φ is then of Gevrey order 1 and its minor is an endlessly continuable function.

Substitution   Let us consider now a *small* power series φ of Gevrey order 1, whose minor φ is endlessly continuable, and denote by $f \in C\{u\}$ a *convergent expansion at the origin*. The power series found by *substituting* φ for the indeterminate variable u in the expansion f is obviously of Gevrey order 1. Let us also analyse the singularities of the iterated convolution products φ^{*n} of φ. Observe here that the successive iterations make some new singularities appear but "further and further away": for all fixed $L > 0$, the sequence $\Omega_L(\varphi^{*n})$ becomes a stationary sequence for some sufficiently large n. It is thus possible to build an endlessly Riemann surface where all the iterated convolution product φ^{*n} are analytic functions. Moreover, one proves ([CNP] [Ph2]) that the Borel transform of $f(\varphi)$ can be extended on this Riemann surface. The following theorem summarises our result.

Theorem 1.1 *When substituting a small power series $^b\varphi$ of Gevrey order 1 with an endlessly continuable minor for the indeterminate variable of a convergent expansion $f \in C\{u\}$, one gets a formal power series $f(^b\varphi)$ of Gevrey order 1 whose minor is an endlessly continuable function.*

1.2.3.   Simple resurgent functions

Go back now to the previous first example and note that the convolution product of two polar singularities gives rise to some polar or logarithmic singularities. More generally, it is easy to see that the convolution product of polar or logarithmic singularities gives the same type of singularities. This observation leads to the following definition.

One says that an endless germ of analytic function at the origin φ has only *simple singularities* if for all paths λ ending at a singular point ω, the analytic continuation $\lambda\varphi$ of φ in a neigbourhood of ω is locally of the form

$$\lambda\varphi(\xi) = \frac{a_\lambda}{2\pi i(\xi - \omega)} + \varphi_\lambda(\xi - \omega)\frac{\log(\xi - \omega)}{2\pi i} + h_\lambda(\xi - \omega),$$

where a_λ is a complex number, φ_λ and h_λ some analytic germs at the origin.

Remark: By the way, the germs φ_λ and h_λ are also endlessly continuable functions with simple singularities.

Definition 1.2 *A simple resurgent function is a formal power series of Gevrey order 1 whose minor is an endlessly continuable function with only simple singularities.*

Theorem 1.2 *The set of simple resurgent functions is a subalgebra of* $C[[x^{-1}]]$, *denoted by* $^+\mathcal{R}(1)$.

For the proof, see [E1] or [CNP].

This algebra is obviously stable under the natural differentiation $\frac{\partial}{\partial x}$. But what about substitution? the Theorem 1.1 shows that if $^b\varphi$ is a small formal power series and f a convergent expansion at the origin, then the Gevrey order 1 expansion $f(^b\varphi)$ has an endlessly continuable minor. Nevertheless, if $^b\varphi$ belongs to $^+\mathcal{R}(1)$, it is not true "à priori" that it is the same for $f(^b\varphi)$ and we should in fact consider a convenient extension of our resurgence algebra. One may prove however (see [CNP]) that this result is true in $^+\mathcal{R}(1)$ for instance for $\exp{^b\varphi}$, $\log{^b\varphi}$, or $(1 - {^b\varphi})^{-1}$. In particular, this last example induces the following corollary.

Corollary 1.3 *The invertible elements of the algebra of simple resurgent functions are those whose residual coefficients vanish – in other words, those which are invertible in* $C[[x^{-1}]]$.

1.2.4.   Elementary resurgent symbols

An *elementary resurgent symbol* is a formal object, a product of a resurgent function φ and an exponential:

$$\dot\varphi^\omega \overset{def}{=} \varphi e^{-x\omega}.$$

The point $\omega \in \mathbf{C}$ is called the *support* of the symbol $\dot{\varphi}^{\omega}$. In particular the resurgent functions are the resurgent symbols with zero support. As the multiplication by an exponential $e^{-x\omega}$ becomes a translation via the Borel transformation, one calls the *minor* of $\dot{\varphi}^{\omega}$ the analytic germ deduced from the minor φ of φ by translation:

$$\dot{\varphi}^{\omega}(\xi) \stackrel{def}{=} \varphi(\xi - \omega).$$

1.3. Summations and connection automorphism

1.3.1. Lateral Borel summations

Let α be a direction in the ξ-plane and φ a resurgent function. Consider the integrals

$$s_{\alpha_{\pm}} \varphi(x) = a_0 + \int_{0_{\alpha_{\pm}}} e^{-x\xi}\varphi(\xi)\,d\xi$$

where a_0 is the residual coefficient and φ the minor of φ. We denote by 0_{α_+} (resp. 0_{α_-}) the endless path starting from the origin to infinity along the direction α and avoiding the possible singularities of φ towards $Arg\,\xi > Arg\,\alpha$ (resp. $Arg\,\xi < Arg\,\alpha$).

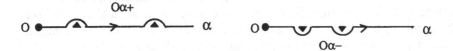

The singularities in the direction α are represented by "nails". The nails come up in the direction where the singularities are seen (don't forget that a singularity may be seen from the left and not from the right and vice versa).

If the rate of growth of φ is of *exponential type* τ at infinity, that is if its modulus is less than $ce^{\tau'|\xi|}$ for all $\tau' > \tau$, then the previous integrals are convergent and define two analytic functions of x in the *sectorial neighbourhood of infinity*, $\Re e(xe^{iArg\alpha}) > \tau$.

If the direction α is not a singular direction for the minor φ, then the integral $s_\alpha \varphi(x) \stackrel{def}{=} s_{\alpha_+} \varphi(x) = s_{\alpha_-} \varphi(x)$ is the *Borel sum* of φ in the direction α. One says alternatively that $s_{\alpha_+} \varphi$ (resp. $s_{\alpha_-} \varphi$) is the *right lateral Borel sum* (resp. *left lateral Borel sum*) of φ in the direction α. So we can define a subalgebra of all *summable resurgent functions*, linked to the algebra of analytic functions of x in some sectorial neigbourhood of infinity by the *homomorphism* s_{α_+} or s_{α_-}.

Remark: In the case when we do not control the growth at infinity, the previous integrals maintain a meaning provided we truncate the path of integration "as far as we want", thus defining analytic functions of x modulo exponential functions "as small as we want", to be more precise *modulo the vanishing exponential functions*, that is the functions of exponential type $\tau = -\infty$. The operators s_{α_+} and s_{α_-} are then called *presummation homomorphisms*.

The *lateral Borel sum of an elementary resurgent symbol* $\dot{\varphi}^\omega = \varphi e^{-x\omega}$ is obviously defined by writing

$$\dot{s}_{\alpha_\pm}\dot{\varphi}^\omega = (s_{\alpha_\pm}\varphi)e^{-x\omega}.$$

Suppose now that A is an arc of the circle connecting the directions in the ξ-plane such that $\alpha \in A$ is the unique singular direction for the elementary resurgent symbol $\dot{\varphi}^\omega$. For $\beta \in A$ such that $Arg\,\beta > Arg\,\alpha$ (resp. $Arg\,\beta < Arg\,\alpha$), we deduce from the Cauchy theorem that the Borel sum $\dot{s}_{\alpha_+}\dot{\varphi}^\omega$ (resp. $\dot{s}_{\alpha_-}\dot{\varphi}^\omega$) coincides with $\dot{s}_\beta\dot{\varphi}^\omega$ on the intersection of the domains $\Re(xe^{\imath Arg\,\alpha}) \gg 0$ and $\Re(xe^{\imath Arg\,\beta}) \gg 0$ and then they are *analytic continuations* the one from the other.

1.3.2.   Connection automorphism and resurgent symbols

Let α be a singular direction for the simple resurgent function φ. In order to compare the left and right Borel sums, assume at the moment that φ (in fact its minor φ) has only one singularity ω along the direction α. The principal part of φ has the following shape, locally in a neigbourhood of ω :

$$\lambda\varphi(\xi) = \frac{a_\omega}{2\pi\imath(\xi-\omega)} + \varphi_\omega(\xi-\omega)\frac{\log(\xi-\omega)}{2\pi\imath} + hol._\omega(\xi-\omega);$$

hence we deduce

$$(s_{\alpha_+} - s_{\alpha_-})\,\varphi(x) = \int_{C_{\omega_\alpha}} e^{-x\xi}\varphi(\xi)\,d\xi$$

$$= -e^{-x\omega}s_\alpha(a_\omega + {}^b\varphi_\omega)$$

$$= \dot{s}_\alpha\{(a_\omega + {}^b\varphi_\omega)e^{-x\omega}\}$$

where C_{ω_α} is the endless path coming from infinity along the direction α, turning around the singularity ω in the negative sense and going back to infinity along the direction α.

Let now $\mathcal{S}_\omega$ be the operator *"hold of singularity"* at ω, defined by

$$\mathcal{S}_\omega : {}^+\mathcal{R}(1) \to {}^+\mathcal{R}(1),\, \varphi \to a_\omega + {}^b\varphi_\omega,$$

and denote by $\dot{\mathcal{S}}_\omega$ the operator

$$\dot{\mathcal{S}}_\omega \overset{def}{=} e^{-x\omega}\mathcal{S}_\omega.$$

The relationship between right and left lateral Borel summations becomes then

$$\dot{s}_{\alpha_+}(1 + \dot{\mathcal{S}}_\omega) = \dot{s}_{\alpha_-} \text{ or alternatively } \dot{s}_{\alpha_-}(1 - \dot{\mathcal{S}}_\omega) = \dot{s}_{\alpha_+}.$$

More generally, let $\underline{\mathcal{S}}_{\alpha_+}$ (resp. $\underline{\mathcal{S}}_{\alpha_-}$) be the operator "*hold of singularity*" in the direction α of the analytic continuations on the right (resp. on the left),

$$\underline{\mathcal{S}}_{\alpha_\pm}\varphi = \sum_{\omega \in]0_\alpha} \dot{\mathcal{S}}_{\omega_\pm}\varphi.$$

(The sum is over the discrete set of the singularities seen from the right – resp. from the left .)

Bending the path of integration as in the picture below and because of the Cauchy theorem, we get

$$\dot{s}_{\alpha_+}(1 + \underline{\mathcal{S}}_{\alpha_+}) = \dot{s}_{\alpha_-} \text{ or alternatively } \dot{s}_{\alpha_-}(1 - \underline{\mathcal{S}}_{\alpha_-}) = \dot{s}_{\alpha_+},$$

a relationship which extends easily to the elementary resurgent symbols.

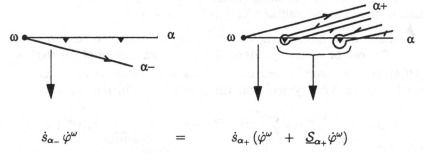

$$\dot{s}_{\alpha_-}\dot{\varphi}^\omega \qquad = \qquad \dot{s}_{\alpha_+}(\dot{\varphi}^\omega + \underline{\mathcal{S}}_{\alpha_+}\dot{\varphi}^\omega)$$

The previous study introduces some new objects, the *resurgent symbols in the direction* α, the discrete sum of elementary resurgent symbols with supports on the half-line $]0_\alpha$,

$$\dot{\varphi} \overset{def}{=} \sum_{\omega \in \Omega(\dot{\varphi})} \dot{\varphi}^\omega.$$

To be more precise, let us denote by A an arc of the circle of the directions in the ξ-plane. We shall say that $\dot{\varphi}$ is a *resurgent symbol in the codirection A* if the sum is over a discrete set $\Omega(\dot{\varphi}) \subset \mathbf{C}$, called the *support of $\dot{\varphi}$*, included in a sectorial neigbourhood of infinity in the codirection A, the intersection of half-planes which cover A.

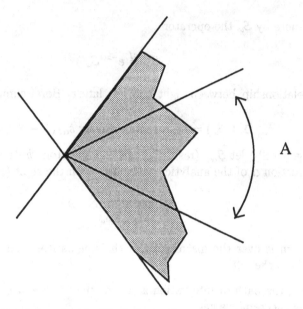

Sectorial neigbourhood of infinity in the codirection A.

Thus defined, the set of resurgent symbols in the codirection A makes up a *multiplicative algebra* denoted by $^+\dot{\mathcal{R}}(1)(A)$. These algebras define a *sheaf* $^+\dot{\mathcal{R}}(1)$ *over the circle of the directions* and the set $^+\dot{\mathcal{R}}_\alpha$ of resurgent symbols in the direction α is the *stalk* of $^+\dot{\mathcal{R}}(1)$ at the point α.

If A is an arc of the circle of the directions on the ξ-plane and $\alpha \in A$, the operator $\underline{S}_{\alpha_\pm}$ can be obviously extended as a homomorphism of the algebra $^+\dot{\mathcal{R}}(1)(A)$. It is as natural to extend the lateral Borel summations homomorphisms $\dot{s}_{\alpha_\pm}$ to resurgent symbols into the codirection A by the following:

$$\text{if } \dot{\varphi} = \sum_{\omega \in \Omega(\dot{\varphi})} \dot{\varphi}^\omega = \sum_{\omega \in \Omega(\dot{\varphi})} \varphi^\omega e^{-x\omega} \text{ then } \dot{s}_{\alpha_\pm} \overset{def}{=} \sum_{\omega \in \Omega(\dot{\varphi})} \dot{s}_{\alpha_\pm} \dot{\varphi}^\omega.$$

Under what conditions does this last integral converge? Note at first that for x large and in order to bound the $e^{-x\omega}$, $\omega \in \Omega(\dot{\varphi})$, we need to restrict x such that $Arg\, x \in \bar{A}$ (complex conjugate of A). One proves in fact ([E1] [CNP]) that this condition implies the convergence of the expansion *modulo the exponentially vanishing functions*, and $\dot{s}_{\alpha_\pm}$ are then *presummation homomorphisms*. We may suppose the summability afterwards by considering the subalgebra of *summable resurgent symbols*.

The relationships between lateral Borel summations

$$\dot{s}_{\alpha_+}(1 + \underline{S}_{\alpha_+}) = \dot{s}_{\alpha_-} \text{ or alternatively } \dot{s}_{\alpha_-}(1 - \underline{S}_{\alpha_-}) = \dot{s}_{\alpha_+}$$

give lastly one of the keys of resurgence, namely the *connection (or Stokes) automorphism*.

Proposition 1.2 *The $^+\dot{\mathcal{R}}(1)(A)$ homomorphism $\underline{S}_\alpha \overset{def}{=} 1 + \underline{S}_{\alpha_+}$ is an automorphism called the connection automorphism in the direction α. Its inverse is given by $\underline{S}_\alpha^{-1} = 1 - \underline{S}_{\alpha_+}$ and moreover we have the relation*

$$\dot{s}_{\alpha_+} \underline{S}_\alpha = \dot{s}_{\alpha_-}.$$

1.4.   Alien derivatives

1.4.1.   Alien differentiations of resurgent symbols

In the same way that the differentiation $\frac{\partial}{\partial x}$ is induced by the automorphism $\mathcal{A} : f(x) \to f(x+1)$ when writing $\mathcal{A} = \exp\frac{\partial}{\partial x}$ (Taylor's formula), the automorphism $\underline{S}_\alpha$ gives rise to a differentiation $\underline{\Delta}_\alpha$ of the algebra $^+\dot{\mathcal{R}}(1)(A)$ by the relation

$$\underline{S}_\alpha \overset{def}{=} \exp\underline{\Delta}_\alpha,$$

or again, with respect to the equality $\underline{S}_\alpha = 1 + \underline{S}_{\alpha_+}$,

$$\underline{\Delta}_\alpha = \log\underline{S}_\alpha = \sum_{n\geq 1} \frac{(-1)^{n-1}}{n}(\underline{S}_{\alpha_+})^n.$$

The operator $\underline{\Delta}_\alpha$ is the *directional differentiation* (in the direction α). Its decomposition in homogenous components gives an infinity of differentiations $\dot{\Delta}_\omega$ or Δ_ω,

$$\underline{\Delta}_\alpha = \sum_{\omega \in]0_\alpha} \dot{\Delta}_\omega \text{ with } \dot{\Delta}_\omega = e^{-x\omega}\Delta_\omega \text{ and,}$$

$$\dot{\Delta}_\omega = \sum_{\substack{\omega_1+\ldots+\omega_n=\omega \\ \omega_i \in]0_\alpha}} \frac{(-1)^{n-1}}{n}(\dot{S}_{\omega_{1_+}})\ldots(\dot{S}_{\omega_{n_+}}).$$

The differentiations Δ_ω act on $^+\mathcal{R}(1)(A)$ and are called *alien differentiations*, while the *dotted alien differentiations* $\dot{\Delta}_\omega$ act on the resurgent symbols of $^+\dot{\mathcal{R}}(1)(A)$.

In order to describe the Stokes phenomena of some resurgent functions or resurgent symbols, it is then equivalent to describe the action of either the connection automorphism or the directional differentiation. The last method comes down to describing the action of the whole alien differentiations. Although appearing to be tedious, this method is often used, as it makes resurgent functions appear most of the time as solutions of some *resurgence equations* which have similarities to ordinary differential equations (see 1.4.3. below).

1.4.2.   Commutation with the natural differentiation

The natural differentiation $\frac{\partial}{\partial x}$ can be naturally extended to $^{+}\dot{\mathcal{R}}(1)(A)$ by setting

$$\partial \Big(\sum_{\omega \in \Omega(\varphi)} \varphi^{\omega} e^{-x\omega} \Big) = \sum_{\omega \in \Omega(\varphi)} (-\omega \varphi^{\omega} + \partial \varphi^{\omega}) e^{-x\omega}.$$

As $\frac{\partial}{\partial x}$ commutes with the lateral Borel summations, it commutes also with the connection automorphism and therefore with the directional differentiation $\underline{\Delta}_{\alpha}$. We deduce that *the dotted alien differentiations $\dot{\Delta}_{\omega}$ commute with the natural differentiation $\frac{\partial}{\partial x}$*. Then the equality $\dot{\Delta}_{\omega} = e^{-x\omega} \Delta_{\omega}$ immediately gives the following rule of commutation,

$$[\Delta_{\omega}, \partial] = -\omega \Delta_{\omega}.$$

1.4.3.   Resurgence equations

As in the usual differential calculus where the objects are often known as solutions of differential equations, we may have to solve in alien differential calculus some *resurgence equations* of the following form:

$$\begin{cases} \Delta_{\omega_i} \varphi = \psi_i & \text{for } \omega_i \in \Omega(\varphi), \\ \Delta_{\omega} \varphi = 0 & \text{for } \omega \notin \Omega(\varphi), \end{cases}$$

where $\Omega(\varphi)$ is a discrete subset of $\mathbf{C} \backslash \{0\}$ and $\psi_i \in {}^{+}\dot{\mathcal{R}}(1)$. We are going to give here, ways to proceed with some simple examples.

Constants of resurgence   Let us consider in $^{+}\mathcal{R}(1)$ the following resurgence equation:

$$\text{for all } \omega \in \mathbf{C} \backslash \{0\}, \ \Delta_{\omega} \varphi = 0.$$

Such a resurgent function φ is called a *constant of resurgence*. The set of constants of resurgence makes up a subalgebra of $^{+}\mathcal{R}(1)$. The minor φ of a constant of resurgence has nevertheless no singularity in the ξ-plane. Moreover, if we assume the summability, the resurgent function φ is Borel summable over all the directions and then defines by summation a holomorphic function in a neigbourhood of infinity. In other words, the formal power series φ is a *convergent expansion*.

Linear system of resurgence equations   Resurgence equations can be often treated by analogy with classical differential calculus. For instance, in the same way that the solutions of a linear system of differential equations are found as linear combinations of an independent system of particular solutions, the solutions of a linear system of resurgence equations are determined by some combinations of particular independent solutions, while the roles of the constant functions are played by the constants of resurgence. Thus the system

$$\begin{cases} \Delta_{\omega_1}\varphi = a(x)\varphi + b(x)\psi, \\ \Delta_{\omega_2}\psi = c(x)\varphi + d(x)\psi, \\ \Delta_\omega\varphi = 0 \text{ for } \omega \neq \omega_1 \text{ and } \Delta_\omega\psi = 0 \text{ for } \omega \neq \omega_2, \end{cases}$$

where a, b, c, d are resurgent functions and ω_1, ω_2 are two nonzero complex numbers, admits as general solution the pair

$$\begin{pmatrix} \varphi \\ \psi \end{pmatrix} = \theta_1 \begin{pmatrix} \varphi_1 \\ \psi_1 \end{pmatrix} + \theta_2 \begin{pmatrix} \varphi_2 \\ \psi_2 \end{pmatrix}$$

where θ_1 and θ_2 are arbitrary constants of resurgence, and where the pairs (φ_1, ψ_1), (φ_2, ψ_2) are two particular solutions such that $\varphi_1\psi_2 - \varphi_2\psi_1$ is invertible.

A simple example of alien calculus   We want to solve here the following resurgence equation in $^+\mathcal{R}(1)$:

$$\begin{cases} \Delta_{n\omega}\varphi = \frac{(-1)^n}{n}\varphi & \text{for } n \in \mathbf{Z}\backslash\{0\}, \\ \Delta_\mu\varphi = 0 & \text{for } \mu \neq \omega, \end{cases}$$

for a fixed $\omega \in \mathbf{C}\backslash\{0\}$.

Denote by $\mathcal{U}^\omega(x)$ the *small (summable) resurgent function* whose minor is equal to $\frac{1}{2\pi i(\xi-\omega)}$. It satisfies the resurgence equation,

$$\begin{cases} \Delta_\omega\mathcal{U}^\omega = 1, \\ \Delta_\mu\mathcal{U}^\omega = 0 & \text{for } \mu \neq \omega. \end{cases}$$

Moreover, as it is a small function, its exponential $\exp\mathcal{U}^\omega(x)$ defines a (summable) resurgent function in $^+\mathcal{R}(1)$ (see 1.2.3) which is a solution of the resurgent equation

$$\begin{cases} \Delta_\omega\exp\mathcal{U}^\omega = \exp\mathcal{U}^\omega, \\ \Delta_\mu\exp\mathcal{U}^\omega = 0 & \text{for } \mu \neq \omega, \end{cases}$$

It is then easy to see that the expansion (convergent summable in $^+\mathcal{R}(1)$)

$$\varphi_0 = \prod_{n \in \mathbf{Z}\backslash\{0\}} \exp(\frac{(-1)^n \mathcal{U}^{n\omega}}{n})$$

is a particular solution of our first resurgence equation. Then the general solution is $\theta(x)\varphi_0$ where θ is an arbitrary constant of resurgence.

Remark: the logarithm of the particular solution φ_0 is a small resurgent function whose minor is

$$B(\log \varphi_0)(\xi) = \frac{1}{2\pi\imath\omega}\left(\frac{1}{y^2} - \frac{\pi}{y\sinh \pi y}\right) \text{ where } y = \frac{\imath\xi}{\omega}.$$

Resurgence monomials More generally, one often uses *resurgence monomial* expansions, to solve resurgence equations, the simplest example is given by the *canonical resurgence monomials*. Roughly speaking, they are small resurgent functions denoted by $\mathcal{U}^{\underline{\omega}}$ where $\underline{\omega} = (\omega_1, \omega_2, \ldots, \omega_n)$, $\omega_i \in \mathbf{C}\backslash\{0\}$, is a multi-index, satisfiying a Δ-stability condition,

$$\begin{cases} \mathcal{U}^{\emptyset} = 1, \\ \Delta_{\omega_1}\mathcal{U}^{\underline{\omega}} = \mathcal{U}^{\underline{\omega}'}, \\ \Delta_{\mu}\mathcal{U}^{\underline{\omega}} = 0 & \text{for } \mu \neq \omega_1, \end{cases}$$

with $\underline{\omega}' = (\omega_2, \ldots, \omega_n)$.

If ω is a simple index, then the term $\mathcal{U}^{\omega}$ has already been described above. The construction of the minor of $\mathcal{U}^{\underline{\omega}}$ is done by iteration with the help of the minor of $\mathcal{U}^{\omega}$ by convolution products [E1], [CNP].

As an application, let us consider the following system,

$$\begin{cases} \Delta_{\omega_1}\varphi = \psi, \\ \Delta_{\omega_2}\psi = \varphi, \\ \Delta_{\omega}\varphi = 0 \text{ for } \omega \neq \omega_1 \text{ and } \Delta_{\omega}\psi = 0 \text{ for } \omega \neq \omega_2, \end{cases}$$

where ω_1 and ω_2 belong to $\mathbf{C}\backslash\{0\}$. Particular solutions can be found by observation. For instance here, by considering φ and ψ as series of $\mathcal{U}^{\underline{\omega}}$ for some multi-index containing only ω_1 and ω_2, and by identification, it is easy to see that the expansions

$$\varphi_1 = \mathcal{U}^{\emptyset} + \mathcal{U}^{\omega_1} + \mathcal{U}^{\omega_1,\omega_2} + \mathcal{U}^{\omega_1,\omega_2,\omega_1} + \mathcal{U}^{\omega_1,\omega_2,\omega_1,\omega_2} + \ldots,$$

$$\psi_1 = \mathcal{U}^{\emptyset} + \mathcal{U}^{\omega_2} + \mathcal{U}^{\omega_2,\omega_1} + \mathcal{U}^{\omega_2,\omega_1,\omega_2} + \mathcal{U}^{\omega_2,\omega_1,\omega_2,\omega_1} + \ldots$$

are formally solutions of the system. We have to confess however that the convergence of the minor is generally hard to prove ...

Note finally that in practice, we often have to prove the resurgence, in order to build resurgence monomials fitted to each particular problem. We shall see such an example in the next chapter. To go further, see [E1].

2.   Equational and quantum resurgence

We intend to study here the resurgence of some particular differential equations. They will be of the form

$$\frac{\partial}{\partial z}\psi = F(z, x, \psi),$$

where x is a *nonzero* complex parameter and F an analytic function to be made precise.

First, we shall analyse the resurgent properties of formal power series with respect to the variable z of integration. Then, starting with an asymptotic solution in x at infinity, we shall introduce the notion of *quantum (or coequational) resurgence*. We shall thus demonstrate a duality between these two types of resurgence.

2.1.   First example

Up to the notation, this first example is the one developed in 1.2.2, namely the Riccati differential equation,

$$\frac{\partial}{\partial z}\psi - x\psi = -\frac{1}{z} - \frac{1}{z^2}\psi^2. \tag{3}$$

We are going to begin with the study of the resurgence with respect to the variable z of integration, in other words, the equational resurgence.

2.1.1.   Equational resurgence

Formal solution   Let $\varphi = \varphi(z, x)$ be the unique power series at infinity, without constant term, which is a formal solution of the equation (3):

$$\varphi(z, x) = x^{-1}z^{-1} - x^{-2}z^{-2} + 2x^{-3}z^{-3} + (-6x^{-4} + x^{-3})z^{-4}$$

$$+(24x^{-5} - 6x^{-4})z^{-5} + \ldots \in \mathbb{C}[x^{-1}][[z^{-1}]].$$

We call it the *formal solution*. We have already described in 1.2.2 a method to analyse its Borel transform: constructing it by successive approximations, we get moreover a precise description of the singularities. Recall here that the formal power series φ is Gevrey order 1 and its minor $\hat{\varphi}$ defines a germ of analytic function at the origin with endless continuation on the universal covering of $\mathbb{C}\backslash\{-x, -2x, -3x, \ldots\}$.

Formal integral The formal solution is far from giving the whole set of formal solutions of equation (3). We get for instance a family of formal solutions adjacent to φ, parametrised by $u \in \mathbf{C}$, by setting

$$\begin{cases} \hat{\varphi}(z,x,u) = \varphi_0(z,x) + u\hat{\varphi}_1(z,x) + u^2\hat{\varphi}_2(z,x) + \dots \\ \qquad \text{and} \\ \frac{\partial}{\partial z}\hat{\varphi}(z,x,u) - x\hat{\varphi}(z,x,u) = -\frac{1}{z} - \frac{1}{z^2}\hat{\varphi}(z,x,u) . \end{cases}$$

The expansion φ_0 is nothing else but the formal solution previously described. On the other hand, when identifying the powers of u in the equation, we obtain for instance for the first terms

$$\begin{cases} \frac{\partial}{\partial z}\hat{\varphi}_1 - x\hat{\varphi}_1 = -\frac{2}{z^2}\varphi_0\hat{\varphi}_1 , \\ \frac{\partial}{\partial z}\hat{\varphi}_2 - x\hat{\varphi}_2 = -\frac{2}{z^2}\varphi_0\hat{\varphi}_2 - \frac{1}{z^2}\hat{\varphi}_1^2 , \\ \qquad \dots \end{cases} \qquad (4)$$

One easily checks that the formal solutions of the first equation are as follows

$$\hat{\varphi}_1(z,x) = e^{xz}\varphi_1(z,x),$$

and we can fix the normalisation of φ_1 by setting $\varphi_1 = 1 + O(z^{-1})$ so that

$$\varphi_1(z,x) = 1 + x^{-1}z^{-2} - \frac{2}{3}x^{-2}z^{-3} + x^{-3}z^{-4}$$

$$+ \frac{2}{5}(-6x^{-4} + x^{-3})z^{-5} + \dots \in \mathbf{C}[x^{-1}][[z^{-1}]].$$

We obtain more generally

Proposition 2.1 *There exists a formal solution,*

$$\hat{\varphi}(z,x,u) = \varphi_0(z,x) + u\hat{\varphi}_1(z,x) + u^2\hat{\varphi}_2(z,x) + \dots,$$

with

$$\hat{\varphi}_m(z,x) = e^{mxz}\varphi_m(z,x), \ \varphi_m(z,x) \in \mathbf{C}[x^{-1}][[z^{-1}]],$$

unique up to the choice of the normalisation of $\varphi_1(z,x)$. It is the formal integral of equation (3).

Endless continuability The formal integral appears to be a "symbol" with support $\{0, -x, -2x, \dots\}$. Is it resurgent? To answer this question, we have to analyse its Borel transform, and for that we are going to study first the formal power series φ_1. We deduce the following from the relation (4):

$$\frac{\partial}{\partial z}\varphi_1 = -\frac{2}{z^2}\varphi_0\varphi_1,$$

or again when writing $\varphi_1 = 1 + {}^b\varphi_1$,

$$\frac{\partial}{\partial z}{}^b\varphi_1 = -\frac{2}{z^2}(\varphi_0 + {}^b\varphi_1).$$

Like the study of the minor of the formal solution, we build the minor $\varphi_1(\zeta, x)$ by successive approximations by considering the expansion with respect to ε,

$${}^b\varphi_1(z, x, \varepsilon) = g_0(z, x) + \varepsilon g_1(z, x) + \varepsilon^2 g_2(z, x) + \ldots$$

a solution of the equation

$$\frac{\partial}{\partial z}{}^b\varphi_1(z, x, \varepsilon) = -\frac{2}{z^2}(\varphi_0(z, x) + \varepsilon {}^b\varphi_1(z, x, \varepsilon)),$$

then we set $\varepsilon = 1$. If we denote by $g_i(\zeta, x)$ the minor of $g_i(z, x)$, we get in this way

$$\begin{cases} \frac{\partial}{\partial z}g_0 = -\frac{2}{z^2}\varphi_0 & \xrightarrow{B} & -\zeta g_0 = -2\zeta * \varphi_0 \\ \frac{\partial}{\partial z}g_{i+1} = -\frac{2}{z^2}g_i & \xrightarrow{B} & -\zeta g_{i+1} = -2\zeta * g_i \quad (i \geq 1). \end{cases}$$

Note here that the convolution product $\zeta * \varphi_0$ defines a germ of an analytic function which vanishes at the origin. We deduce that $g_0(\zeta, x)$ is a regular function at the origin and, by iteration, it is the same for all $g_i(\zeta, x)$. When recalling that $\varphi_0(\zeta, x)$ is a germ of an analytic function at the origin with analytic continuation on the universal covering of $\mathbf{C}\backslash\{-x, -2x, -3x, \ldots\}$, it is then obvious that the $g_i(\zeta, x)$ are germs of analytic function at the origin with analytic continuation on the universal covering of $\mathbf{C}\backslash\{0, -x, -2x, -3x, \ldots\}$. We shall admit here the convergence of φ_1 which can be proved by an argument of bounding expansion (see [CNP]). The analysis of the φ_m can be done in the same manner. It opens with the following result:

Proposition 2.2 *For $m = 0, 1, 2, \ldots$, the expansion $\varphi_m(\zeta, x)$ defines a germ of an analytic function at the origin which extends to the universal covering of $\mathbf{C}\backslash\{(m-1)x, (m-2)x, \ldots, 0, -x, -2x, \ldots\}$.*

Remark: the bounding expansion method allows us to prove that in every sector avoiding the singularities, the principal determination of $\varphi_m(\zeta, x)$ is of exponential type at infinity, precisely,

$$|\varphi_m(\zeta, x)| \leq C(x)a(x)^m e^{\tau(x)|\zeta|}, \quad C(x), a(x), \tau(x) > 0,$$

which enables the analysis of the summability of the formal integral.

Resurgence equations The formal integral is then Borel summable in all the directions of the ζ-plane except those given by the argument of x or counterwise to it. In order to make precise the summations, we have now to analyse the action of the connection automorphism for these singular directions. Knowing the position of the (possible) singularities of the formal integral, it is better to consider the action of the alien differentiations Δ_ω for ω in $x\mathbf{Z}\backslash\{0\}$. More exactly, we are going to use the dotted alien differentiations $\dot\Delta_\omega$ which commute (see 1.4.2) with the natural differentiation $\frac{\partial}{\partial z}$. From the equality

$$(\frac{\partial}{\partial z} - x)\dot\varphi = -\frac{1}{z} - \frac{1}{z^2}\dot\varphi^2,$$

we get

$$(\frac{\partial}{\partial z} - x)\dot\Delta_\omega\dot\varphi = -\frac{1}{z^2}\dot\varphi\dot\Delta_\omega\dot\varphi.$$

But in the same way,

$$(\frac{\partial}{\partial z} - x)\frac{\partial}{\partial u}\dot\varphi = -\frac{1}{z^2}\dot\varphi\frac{\partial}{\partial u}\dot\varphi.$$

Since the symbols $\dot\Delta_\omega\dot\varphi$ and $\frac{\partial}{\partial u}\dot\varphi$ are solutions of the same homogenous equation of order 1 in the variable z, they are then proportional:

$$\dot\Delta_\omega\dot\varphi = a_\omega(x,u)\frac{\partial}{\partial u}\dot\varphi.$$

When expanding the formal integral in the previous equality and comparing the same weight components, it is easy to make precise the coefficient of proportionality and thus obtain the following proposition:

Proposition 2.3 *The formal integral $\dot\varphi$ satisfies the "bridge equation":*

$$\begin{cases} \text{if } \omega = kx, \text{ with } k = -1, 1, 2, \ldots, \text{ then} \\ \qquad \dot\Delta_\omega\dot\varphi = A_k(x)u^{(1+k)}\frac{\partial}{\partial u}\dot\varphi \; ; \\ \text{otherwise,} \\ \qquad \dot\Delta_\omega\dot\varphi = 0. \end{cases}$$

Comments:

• This equation creates a "bridge" between alien differential calculus and ordinary differential calculus, hence the obvious name.

• One can deduce from the proposition above the resurgent equations of each elementary symbol of the formal integral. We get for instance for the formal solution,

$$\begin{cases} \qquad \Delta_{-x}\varphi_0 = A_{-1}(x)\varphi_1, \\ \text{otherwise,} \\ \qquad \Delta_\omega\varphi_0 = 0. \end{cases}$$

• For all fixed nonzero x, the resurgent coefficients $A_k(x)$ are well determined complex numbers, eventually vanishing. One can prove that they constitute a *complete set of holomorphic invariants for the analytic conjugation*. To go further, see [E1], [E2], [CNP], [F].

Confluence   Let us now look at the parameter x as a complex variable. We recall that the power series $\varphi_m(\zeta, x)$ appear as formal expansions in z^{-1}, with *polynomial coefficients in x^{-1}*. Their minors are then analytic in x on $\mathbf{C}\backslash\{0\}$, hence we deduce that the resurgent coefficients $A_k(x)$ define some analytic functions in x on $\mathbf{C}\backslash\{0\}$.

What happens in a neighbourhood of zero? For x going to the origin, as the singular lattice $x\mathbf{Z}\backslash\{0\}$ includes the homothetic factor x, we observe a *confluence* of the singularities to zero and z ceases to be a good resurgent variable. This suggests making the following change of variable :

$$z \rightarrow y = xz.$$

When writing $\varphi_0(z, x) = \phi_0(y, x)$, the formal solution becomes,

$$\phi_0(y, x) = y^{-1} - y^{-2} + 2y^{-3} + (-6 + x)y^{-4} + (24 - 6x)y^{-5} + \ldots \in \mathbf{C}[x][[y^{-1}]].$$

More generally, the $\phi_m(y, x) = \varphi_m(z, x)$ appear now as formal power series in y^{-1} with *polynomial coefficients in x*. This can be easily explained for equation (3) becomes:

$$\text{if } \Phi(y, x) = \psi(z, x),$$

$$\frac{\partial}{\partial y}\Phi - \Phi = -\frac{1}{y} - \frac{x}{y^2}\Phi^2. \tag{5}$$

Now the formal integral,

$$\dot{\phi}(y, x, u) = \phi_0(y, x) + u\dot{\phi}_1(y, x) + u^2\dot{\phi}_2(y, x) + \ldots,$$

has the lattice $\mathbf{Z}\backslash\{0\}$ as a singular set while the resurgence equations remain the same as previously[1], namely

$$\begin{cases} \text{If } \omega = k, \text{ with } k = -1, 1, 2, \ldots, \text{ then} \\ \quad \dot{\Delta}_\omega\dot{\phi} = A_k(x)u^{(1+k)}\frac{\partial}{\partial u}\dot{\phi} \ ; \\ \text{otherwise,} \\ \quad \dot{\Delta}_\omega\dot{\phi} = 0. \end{cases}$$

---

[1]If S_ν denotes the automorphism of the algebra $^+\mathcal{R}(1)$ defined by

$$S_\nu : \varphi(z) \rightarrow \varphi(\nu^{-1}z),$$

one easily proves [E1] the following rule of commutation:

$$\Delta_\omega S_\nu = S_\nu \Delta_{\omega\nu^{-1}}.$$

Thus, as on the one hand the resurgent coefficients are preserved and, on the other hand, the resurgent functions $\phi_m(y, x)$ are holomorphic functions in x, we deduce that the resurgent coefficients $A_k(x)$ have analytic continuations in a neighbourhood of the origin, and therefore are *entire functions of x*.

To end 2.1.1, remark now that for $x = 0$, equation (5) is nothing else but the Euler equation which has been studied at the very beginning. We know that the formal solution has -1 as sole singularity and that the formal integral can be written as

$$\dot{\phi}(y, 0, u) = \phi_0(y, 0) + u e^y.$$

Moreover,

$$(s_{(\pi)_+} - s_{(\pi)_-})\phi_0(y, 0) = -2\pi\imath e^y.$$

This last equality can be rewritten as

$$s_{(\pi)_+}(\phi_0 + 2\pi\imath e^y) = s_{(\pi)_-}\phi_0,$$

hence we deduce

$$\underline{S}_{(\pi)}\phi_0 = \phi_0 + 2\pi\imath e^y$$

and then

$$\Delta_{-1}\phi_0(y, 0) = 2\pi\imath.$$

To summarise, the resurgent coefficients are entire functions of x and for $k \geq 1$ the coefficients $A_k(x)$ vanish for $x = 0$ while $A_{-1}(0) = 2\pi\imath$.

Conclusion   To fix ideas, assume that x is a real positive number. We have seen previously that the formal solution φ_0 defines a small formal power series of Gevrey order 1 whose minor is endlessly continuable with a singular lattice in the real negative direction. The formal expansion φ_0 is Borel summable and defines by summation an analytic function of z in the half-plane $\Re e(z) > \tau$. By rotating the direction of summation in the sector $-\pi < Arg\,\zeta < +\pi$, one extends $s_{(0)}\varphi_0$ into a domain included in $\frac{-3\pi}{2} < Arg\,z < \frac{+3\pi}{2}$.

The nearby solutions are got by summing the formal integral. As the real positive direction is singular for $\dot{\varphi}$, we have to introduce the right and left lateral Borel summations, thus defining two adjacent families of solutions parametrised by u,

$$s_{(0)_\pm}\dot{\varphi}(z, x, u) = \sum_{m \geq 0} (s_{(0)_\pm}\varphi_m)(u e^{xz})^m.$$

These expansions are convergent for $\Re e(z)$ "as large as $|u|$ is sufficiently small". The comparison of these two families of solutions is given by the connection formula,

$$\underline{\Delta}_{(0)}\dot{\varphi} = \sum_{\substack{\omega=k x \\ k>0}} \dot{\Delta}_\omega \dot{\varphi} = \sum_{\substack{\omega=k x \\ k>0}} (A_k(x)u^{(1+k)})\frac{\partial}{\partial u}\dot{\varphi} = A(x,u)\frac{\partial}{\partial u}\dot{\varphi}.$$

Then we deduce

$$\underline{S}_{(0)}\dot{\varphi} = \exp(A(x,u)\frac{\partial}{\partial u})\dot{\varphi}.$$

One proves (see for instance [CNP]) that these connection formulae are *summable*, that is the relations between $s_{(0)_+}\dot{\varphi}$ and $s_{(0)_-}\dot{\varphi}$ are given by some functions of z – and not modulo ... .

The sums $s_{(0)_+}\dot{\varphi}$ and $s_{(0)_-}\dot{\varphi}$ are called *repulsive solutions*. The real trajectories of these solutions start to go along the trajectory of the Borel sum of the formal solution and then go away exponentially quickly (if all coefficients A_k do not vanish). By summing in the direction (π) or $(-\pi)$, we should get the so-called *attractive solutions*.

2.1.2.   Quantum resurgence

Formal solution and formal integral   We consider now x as a "large" complex parameter. By a rearrangement of the terms, we can see the formal solution of equation (3) as a formal expansion in the powers of x^{-1},

$$\varphi_0(x,z) = z^{-1}x^{-1} - z^{-2}x^{-2} + (2z^{-3} + z^{-4})x^{-3}$$

$$+(-4z^{-3} - 6z^{-4} + 2z^{-5})x^{-4} + \ldots \in C[z^{-1}][[x^{-1}]].$$

In the same way, we can consider the formal integral

$$\dot{\varphi}(x,z,u) = \varphi_0(x,z) + u\dot{\varphi}_1(x,z) + u^2\dot{\varphi}_2(x,z) + \ldots,$$

with

$$\dot{\varphi}_m(x,z) = e^{mzx}\varphi_m(x,z), \quad \varphi_m(x,z) \in C[z^{-1}][[x^{-1}]],$$

as a symbol with support $\{0, -z, -2z, -3z, \ldots\}$, where $\varphi_m(x,z)$ are formal power series in x^{-1} with polynomial coefficients in z^{-1}.

Endless continuability   We want now to prove the resurgence in x. Here again, we are going to build the minors of the $\varphi_m(x,z)$ by successive approximations and we begin first with the minor of the small power series $\varphi_0(x,z)$. We introduce a "small number" ε, and let

$$\varphi_0(x,z,\varepsilon) = g_0(x,z) + \varepsilon g_1(x,z) + \varepsilon^2 g_2(x,z) + \ldots$$

be a solution of the equation

$$\frac{\partial}{\partial z}\varphi_0(x,z,\varepsilon) - x\varphi_0(x,z,\varepsilon) = -\frac{1}{z} - \frac{\varepsilon^2}{z^2}\varphi_0^2(x,z,\varepsilon).$$

Recall that, if we denote by $\mathcal{D}$ the operator $(\frac{\partial}{\partial z} - x)$, we get

$$\mathcal{D}g_0 = -z^{-1}\ ,\quad \mathcal{D}g_1 = 0\ ,\quad \mathcal{D}g_2 = -z^{-2}g_0^2\ ,$$

$$\mathcal{D}g_3 = -2z^{-2}g_0g_1\ ,\quad \mathcal{D}g_4 = -z^{-2}g_1^2 - 2z^{-2}g_0g_2,$$

$$\cdots$$

Note now that if $f(x)$ denotes a *small* resurgent function and $f(\xi)$ its minor, then

$$xf(x) \overset{B}{\to} (\delta(\xi) + \frac{\partial}{\partial\xi})f(\xi). \tag{6}$$

Moreover, we require the minor $\varphi_0(\xi,z)$ to satisfy the following condition at infinity:

for z in a neigbourhood of infinity, $\varphi_0(\xi,z) = O(z^{-1})$.

The first equation thus becomes

$$(\frac{\partial}{\partial z} - \delta(\xi) - \frac{\partial}{\partial\xi})g_0(\xi,z) = -\delta(\xi)z^{-1}.$$

By identification of the $\delta(\xi)$ terms, we deduce

$$g_0(0,z) = -z^{-1},$$

and we have the equality

$$(\frac{\partial}{\partial z} - \frac{\partial}{\partial\xi})g_0(\xi,z) = 0.$$

Then we deduce

$$g_0(\xi,z) = -(z+\xi)^{-1}.$$

The second equation obviously gives

$$g_1(\xi,z) = 0;$$

meanwhile the third equation becomes

$$(\frac{\partial}{\partial z} - \delta(\xi) - \frac{\partial}{\partial\xi})g_2(\xi,z) = -z^{-2}g_0 * g_0 \overset{def}{=} G_2(\xi,z).$$

This equation can be rewritten as

$$g_2(0, z) = 0 \text{ and } (\frac{\partial}{\partial z} - \frac{\partial}{\partial \xi})g_2(\xi, z) = G_2(\xi, z),$$

which gives

$$g_2(\xi, z) = -\int_0^\xi G_2(z + \xi - \xi_1, \xi_1)\, d\xi_1.$$

We shall find in the same manner,

$$g_3(\xi, z) = 0,$$

$$g_4(\xi, z) = -\int_0^\xi G_4(z + \xi - \xi_1, \xi_1)\, d\xi_1,$$

where $G_4(\xi, z) \overset{def}{=} -2z^{-2}g_0 * g_2,$

$$\cdots$$

We want now to sketch the analysis of the endless continuability of these $g_k(\xi, z)$. Assume that the variable is fixed and nonzero. The term $g_0(\xi, z)$ admits a pole at $\xi = -z$, so the function $G_2(\xi, z)$ is analytically ramified on $\mathbb{C}\backslash\{-z, -2z\}$. The analysis of $g_2(\xi, z)$ requires a little more work, the difficulties being of the same type as the study of a convolution product. We shall assume here that $g_2(\xi, z)$ is again an analytically ramified function on $\mathbb{C}\backslash\{-z, -2z\}$. We deduce that the analytic germ $G_4(\xi, z)$ can be extended to the universal covering of $\mathbb{C}\backslash\{-z, -2z, -3z, -4z\}$ ... and one may prove finally that the formal solution $\varphi_0(x, z)$ is a small power series of Gevrey order 1 whose minor can be extended to the universal covering of $\mathbb{C}\backslash\{-z, -2z, -3z, \ldots\}$. The minors of every different power series $\varphi_m(x, z)$ can be treated by the same process, which allows us to establish the following proposition, to be compared with the one for the equational resurgence.

Proposition 2.4 *For $m = 0, 1, 2, \ldots$, the expansion $\varphi_m(\xi, z)$ defines a germ of an analytic function at the origin which extends to the universal covering of $\mathbb{C}\backslash\{(m-1)z, (m-2)z, \ldots, 0, -z, -2z, \ldots\}$.*

Remark: Here again, one can prove that in every sector avoiding the singularities, the principal determination of $\varphi_m(\xi, z)$ is of exponential type at infinity, precisely,

$$|\varphi_m(\xi, z)| \leq C(z)a(z)^m e^{\tau(z)|\xi|}, \ C(z), a(z), \tau(z) > 0,$$

which makes possible the analysis of the summability of the formal integral.

Resurgence equation　The previous result demonstrates a *duality* between equational and quantum resurgence, which is much underlined by the resurgence equations; the following proposition can be proved by tracing what has been done for the equational resurgence, as x is a constant of resurgence:

Proposition 2.5 *The formal integral $\dot{\varphi}$ satisfies the "bridge equation":*

$$\begin{cases} \text{if } \omega = kz, \text{ with } k = -1, 1, 2, \ldots, \text{ then} \\ \qquad \dot{\Delta}_\omega \dot{\varphi} = B_k(x) u^{(1+k)} \frac{\partial}{\partial u} \dot{\varphi} ; \\ \text{otherwise,} \\ \qquad \dot{\Delta}_\omega \dot{\varphi} = 0. \end{cases}$$

The resurgent coefficients $B_k(x)$ are here *resurgent functions in x*. Moreover, as the Borel transform of the formal integral has a lattice of singularities depending on z, we deduce that the $B_k(x)$ are in fact *constants of resurgence*.

To end this section, note that the previous results can be easily generalised to the following equations:

$$\frac{\partial}{\partial z} \psi - x\psi = \sum_{n \geq -1} a_n(z)\psi^n,$$

where the $a_n(z)$ are endlessly continuable entire expansions. For equational resurgence, see [E1], [E2], [F] and [CNP]. For quantum resurgence and more about the duality with equational resurgence, see [E2].

2.2.　Second example

We are going now to be interested in the following Riccati equations:

$$\frac{\partial}{\partial q}\Psi = \Psi^2 - x^2 W(q) = F(q, x, \Psi), \tag{7}$$

where $W(q) \in \mathbf{C}[q]$ is a unitary polynomial of degree m, while x is a nonzero complex parameter.

We begin by following the framework of the preceding section and so we look first for a formal solution. Writing,

$$\frac{\partial}{\partial q}\Psi = (\Psi - xW^{\frac{1}{2}})(\Psi + xW^{\frac{1}{2}}),$$

we observe that by writing $\frac{\partial}{\partial q}W(q) = O(q^{-1})W(q)$ – to obtain the following formal solution, with the choice of the determination of $W^{\frac{1}{2}}$:

$$\psi_0(q,x) = xW^{\frac{1}{2}}(q) + \frac{m}{4q} + O(q^{\frac{-3}{2}}).$$

We get also a family of formal solutions parametrised by $u \in \mathbf{C}$, adjacent to the formal solution, by setting,

$$\begin{cases} \dot{\psi}(q,x,u) = \psi_0(q,x) + u\dot{\psi}_1(q,x) + u^2\dot{\psi}_2(q,x) + \ldots \\ \qquad\qquad \text{such that} \\ \frac{\partial}{\partial q}\dot{\psi}(q,x,u) = F(q,x,\dot{\psi}(q,x,u)). \end{cases}$$

For instance, the first terms have to satisfy

$$\begin{cases} \frac{\partial}{\partial q}\dot{\psi}_1 = \dot{\psi}_1\frac{\partial}{\partial\psi}F(q,x,\psi_0), \\ \frac{\partial}{\partial q}\dot{\psi}_2 = \dot{\psi}_2\frac{\partial}{\partial\psi}F(q,x,\psi_0) + \frac{\dot{\psi}_1^2}{2}\frac{\partial^2}{\partial\psi^2}F(q,x,\psi_0), \\ \qquad\qquad \ldots \end{cases}$$

In particular as

$$\frac{\partial}{\partial\psi}F(q,x,\psi_0) = 2\psi_0 = 2xW^{\frac{1}{2}}(q) + \frac{m}{2q} + O(q^{\frac{-3}{2}}),$$

when setting

$$z(q) = \int_{q^0}^{q} 2W^{\frac{1}{2}}(q')\,dq',$$

we can write $\dot{\psi}_1$ as

$$\dot{\psi}_1(q,x) = q^{\frac{m}{2}}e^{xz(q)}\psi_1,$$

and we shall fix the normalization by taking

$$\psi_1(q,x) = 1 + O(q^{\frac{-1}{2}}) \in \mathbf{C}[[q^{\frac{-1}{2}}]].$$

More generally, we obtain the formal integral of equation (6),

$$\begin{cases} \dot{\psi}(q,x,u) = \psi_0(q,x) + u\dot{\psi}_1(q,x) + u^2\dot{\psi}_2(q,x) + \ldots \\ \qquad\qquad \text{with} \\ \dot{\psi}_n(q,x,u) = (q^{\frac{m}{2}}e^{xz(q)})^n\psi_n(q,x), \quad \psi_n(q,x) \in \mathbf{C}((q^{\frac{-1}{2}})), \end{cases}$$

unique up to the choice of the normalization of $\psi_n(q,x)$.

The formal integral then appears as a (resurgent?) symbol in the variable $z(q)$, with support $\{0, -x, -2x, \ldots\}$, which allows us to think that $z(q)$ is

a *good resurgence variable.* One may prove that, by the change of variable
$q \to z(q)$ and a change of function if necessary, equation (7) can be transformed
into a differential equation similar to the one of subsection 2.1, see [B], [Di],
[CNP].

Actually, this method would be awkward for our case. Indeed, we remark
here that the change of function,

$$\Psi = -\frac{\partial}{\partial q} \log \Phi, \qquad (8)$$

transforms the Riccati equation (7) into a linear differential equation, the
stationary unidimensional Schrödinger equation,

$$\frac{\partial^2}{\partial q^2} \Phi = x^2 W(q) \Phi. \qquad (9)$$

We are going to study the resurgence of this last equation, leaving the reader
to interpret the results for the Riccati equation (7).

2.2.1.   Equational resurgence

Formal solution and formal integral   The formal expansion $\psi_0(q, x)$ becomes by the change of function (8) the formal solution

$$\Phi_0(q, x) = c_-(q, x) e^{-\frac{zx(q)}{2}} \text{ or } c_+(q, x) e^{+\frac{zx(q)}{2}}$$

depending on the choice of a determination of the square root of $W(q)$. Then
we expect that the formal integral of the linear equation (9) can be written as
a linear combination of these two *linearly independent terms.* On the other
hand, the previous analysis suggests the change of variable,

$$q \to z(q) = \int_{q_0}^{q} 2W^{\frac{1}{2}}(q') \, dq'.$$

Equation (9) becomes

$$\left(\frac{\partial^2}{\partial z^2} - 2H(z)\frac{\partial}{\partial z} - \frac{x^2}{4}\right)\Phi = 0, \qquad (10)$$

where, if $q(z)$ denotes the inverse (multivalued) function of $z(q)$,

$$H(z) = \frac{1}{2}\frac{\partial}{\partial z} \log\left(\frac{\partial}{\partial z} q(z)\right).$$

One can prove, thanks to the inverse function theorem, that H is an analytic multivalued function at infinity in z. To be more precise,

$$H(z) = -\frac{m}{2(m+2)z}(1 + O(z^{\frac{-2}{m+2}})) \in z^{-1}\mathbf{C}\{z^{\frac{-2}{m+2}}\}.$$

The change of function

$$\Phi = (\frac{\partial}{\partial z}q(z))^{\frac{1}{2}}\phi$$

allows us furthermore to get rid of the term $\frac{\partial}{\partial z}$ in equation (10) and gives

$$(\frac{\partial^2}{\partial z^2} - \frac{x^2}{4} - H^2 + (\frac{\partial}{\partial z}H))\phi = 0. \tag{11}$$

We now write the formal integral of the Schrödinger equation (10), describing the general formal solution. As this equation is linear of order 2, the formal integral can be written as

$$\dot{\Phi}(z,x) = v_+(x)(\frac{\partial}{\partial z}q(z))^{\frac{1}{2}}\varphi_+(z,x)e^{+\frac{zx}{2}} + v_-(x)(\frac{\partial}{\partial z}q(z))^{\frac{1}{2}}\varphi_-(z,x)e^{-\frac{zx}{2}},$$

where $\varphi_\pm(z,x)$ are formal power series in $\mathbf{C}\{x^{-1}\}[[z^{\frac{-2}{m+2}}]]$, which can be normalised by setting

$$\varphi_\pm(z,x) = 1 + O(z^{\frac{-2}{m+2}}),$$

and where $v_\pm(x)$ are arbitrary functions of x.

Endless continuability   The problem of the resurgence of these formal expansions needs some preliminaries. As the power series $\varphi_\pm(z,x)$ are expansions with respect to fractional powers of z^{-1}, we leave the frame of the algebra of power series of Gevrey order 1. Nevertheless, it is possible to widen the Borel transform to fractional powers of z^{-1} thanks to the identity,

$$\text{for } t > -1, \quad z^{-t-1} = \int_0^{+\infty} e^{-z\zeta}\frac{\zeta^t\Gamma(t+1)}{\Gamma(t+1)}d\zeta,$$

and thus to define the *minor* $\varphi_\pm$ of such a $\varphi_\pm$, as the *small power series* remaining whose residual coefficients vanish. As a rule, this minor, if it converges, is no longer a germ of analytic function at the origin but a *ramified* – or *sectorial* – germ at the origin. The notion of *endless continuability* can be naturally widened to these ramified germs and one may adapt the construction of section 1.   But we shall have to keep in mind that the alien differentiations must be indexed by some ω belonging to the universal covering of $\mathbf{C}\backslash\{0\}$. For more details, see [E1] or [CNP].

In order to prove the resurgence in z of the power series $\varphi_\pm(z,x)$, we are going to build them as expansions of *resurgence monomials*, and for that, we have first to make precise their differential equations.

By the change of function

$$\phi = \varphi_\pm e^{\pm\frac{zx}{2}},$$

we deduce from equation (11):

$$(\frac{\partial^2}{\partial z^2} \pm x\frac{\partial}{\partial z} - H^2 + (\frac{\partial}{\partial z}H))\varphi_\pm = 0, \tag{12}$$

which is respectively equivalent to the system

$$\begin{cases} \varphi_+(z,x) = \varphi_{++}(z,x) + \varphi_{+-}(z,x) \\ \qquad \text{such that} \\ \frac{\partial}{\partial z}\varphi_{+-} = -H\varphi_{++} \text{ and } (\frac{\partial}{\partial z}+x)\varphi_{++} = -H\varphi_{+-}, \end{cases}$$

and[2]

$$\begin{cases} \varphi_-(z,x) = \varphi_{-+}(z,x) + \varphi_{--}(z,x) \\ \qquad \text{such that} \\ \frac{\partial}{\partial z}\varphi_{-+} = -H\varphi_{--} \text{ and } (\frac{\partial}{\partial z}-x)\varphi_{--} = -H\varphi_{-+}. \end{cases}$$

Introduce now the resurgent monomials $\mathcal{W}^\bullet$, defined by iteration by

$$\begin{cases} \mathcal{W}^\emptyset = 1 \\ \text{and} \\ (\frac{\partial}{\partial z}+|\underline{X}|)\mathcal{W}^{\underline{X}} = -H\mathcal{W}^{\underline{X}'} \end{cases} \tag{13}$$

where $\underline{X} = \underbrace{(\omega_1,\omega_2,\ldots,\omega_n)}_{n}$, $\underline{X}' = \underbrace{(\omega_1,\omega_2,\ldots,\omega_{(n-1)})}_{n-1}$

and $|\underline{X}| = \omega_1 + \omega_2 + \ldots + \omega_n$.

Then it is easy to check that the expansion

$$\mathcal{W}^\emptyset + \mathcal{W}^{(x)}(z,x) + \mathcal{W}^{(x,-x)}(z,x) + \ldots$$

$$+\mathcal{W}^{\overbrace{(x,-x,\ldots,x)}^{2p-1}}(z,x) + \mathcal{W}^{\overbrace{(x,-x,\ldots,x,-x)}^{2p}}(z,x) + \ldots \tag{14}$$

and, respectively,

$$\mathcal{W}^\emptyset + \mathcal{W}^{(-x)}(z,x) + \mathcal{W}^{(-x,x)}(z,x) + \ldots$$

---

[2]Actually, the choice of the normalization of $\varphi_\pm(z,x)$ implies that these two formal power series are related by $\varphi_-(z,x) = \varphi_-(z,-x)$.

$$+\mathcal{W}\overbrace{(-x,x,\dots,-x)}^{2p-1}(z,x) + \mathcal{W}\overbrace{(-x,x,\dots,-x,x)}^{2p}(z,x) + \dots,$$

is a formal solution of equation (12).

We require furthermore that the $\mathcal{W}^{\underline{X}}$ be some *small* resurgent functions, which implies the uniqueness of the solutions. Note at this point that the function $H(z)$ is a *small constant* of resurgence for belonging to $z^{-1}\mathbf{C}\{z^{\frac{-2}{m+2}}\}$.

If $\mathcal{W}\overbrace{(+,-,\dots,\varepsilon)}^{n}(\zeta,x)$ denotes the minor of $\mathcal{W}\overbrace{(x,-x,\dots,\varepsilon x)}^{n}(z,x)$ and $H(\zeta)$ the minor of $H(z)$, then we get from (13) for instance for the first terms:

$$\mathcal{W}^{(+)} = (\zeta - x)^{-1}H \;,\quad \mathcal{W}^{(+,-)} = \zeta^{-1}(H * \mathcal{W}^{(+)}) \;,$$

$$\mathcal{W}^{(+,-,+)} = (\zeta - x)^{-1}(H * \mathcal{W}^{(+,-)}) \;,$$

$$\dots$$

In particular, it is then obvious that the $\mathcal{W}^{(+,-,\dots,\varepsilon)}(\zeta,x)$ are sectorial germs at the origin with analytic continuation on the universal covering of $\mathbf{C}\backslash\{0,x\}$.

Lastly, one proves by an argument of bounding expansion (see [CNP]) that the Borel transform of the formal expansion (14), which is nothing else but the Borel transform of the formal power series φ_+, defines a sectorial germ at the origin with analytic continuation on the universal covering of $\mathbf{C}\backslash\{0,x\}$. The same should be done for the Borel transform of the formal power series φ_- and we leave it to the reader.

To summarise, we get

Proposition 2.6 *The formal integral of the Schrödinger equation (9),*

$$\dot{\Phi}(z,x) = v_+(x)(\frac{\partial}{\partial z}q(z))^{\frac{1}{2}}\varphi_+(z,x)e^{+\frac{zz}{2}} + v_-(x)(\frac{\partial}{\partial z}q(z))^{\frac{1}{2}}\varphi_-(z,x)e^{-\frac{zz}{2}},$$

is a resurgent (summable) symbol with respect to z. The minors of the resurgent functions φ_+ and φ_- are analytic multivalued on $\mathbf{C}\backslash\{0,x\}$ and $\mathbf{C}\backslash\{0,-x\}$ respectively.

Resurgence equations The resurgence equations of $\varphi_\pm$ are easy to get thanks to linearity: as the dotted alien differentiations $\dot{\Delta}_\omega$ commute with the natural differentiation $\frac{\partial}{\partial z}$, then the resurgent symbol $\dot{\Delta}_\omega\dot{\Phi}$ is still a solution of the Schrödinger equation (10). On the other hand, as the differentiation

$\overset{\bullet}{\Delta}_\omega$ translates by ω the support of a symbol, we thus get by identification of the supports

$$\begin{cases} \Delta_\omega \varphi_+ = S_\omega(x)\varphi_-, & \text{for } \omega \text{ "over" } x, \\ \Delta_\omega \varphi_- = S_\omega(x)\varphi_+, & \text{for } \omega \text{ "over" } -x, \\ \Delta_\omega \varphi_\pm = 0 & \text{in the other cases.} \end{cases}$$

Moreover, as the resurgent functions $\varphi_\pm$ are formal expansions with respect to powers of $z^{\frac{-2}{m+2}}$, then the alien differentiations are actually indexed by the ω "over" x (resp. $-x$) on the Riemann surface of $z^{\frac{2}{m+2}}$. For all nonzero x, $S_\omega(x)$ are well determined complex numbers and, as the power series $\varphi_\pm$ are analytic (maybe multivalued) in x on $\mathbf{C}\backslash\{0\}$, then $S_\omega(x)$ are also analytic (maybe multivalued) on $\mathbf{C}\backslash\{0\}$. Identically, following what has been done in section 2.1 for the first example, by the change of variable,

$$z \longrightarrow y = xz,$$

the power series $\Upsilon_\pm(y,x) = \varphi_\pm(z,x)$ become formal expansions belonging to $\mathbf{C}\{x^{\frac{2}{m+2}}\}[[z^{\frac{-2}{m+2}}]]$, solutions of

$$\left(\frac{\partial^2}{\partial y^2} \pm \frac{\partial}{\partial y} - H^2(\frac{y}{x}) + H'(\frac{y}{x})\right)\Upsilon_\pm = 0$$

where $H' = \frac{\partial}{\partial z}H$. Hence we deduce that the resurgence coefficients $S_\omega(x)$ belong to $\mathbf{C}\{x^{\frac{2}{m+2}}\}$. The following proposition summarises the previous result:

Proposition 2.7 *The resurgent functions $\varphi_\pm$ satisfy the following resurgence equation: if $\nu = m+2$ for m odd and $\nu = \frac{m+2}{2}$ for m even, then*

$$\begin{cases} \Delta_{x_i}\varphi_+ = S_i(x)\varphi_-, & \text{for } i = 2, 4, \ldots, \nu, \\ \Delta_{-x_i}\varphi_- = S_i(x)\varphi_+, & \text{for } i = 1, 3, \ldots, \nu - 1, \\ \Delta_\omega \varphi_\pm = 0 & \text{in the other cases,} \end{cases}$$

where $\{x_2, \ldots, x_\nu\}$ (resp. $\{-x_1, \ldots, -x_{\nu-1}\}$) are the points on the Riemann surface of $x^{\frac{2}{m+2}}$ over x (resp. $-x$). The resurgent coefficients $S_i(x)$ are the Sibuya coefficients which belong to $\mathbf{C}\{x^{\frac{2}{m+2}}\}$.

We refer the reader to [E2] or [CNP] for more information about equational resurgence of the Schrödinger equation. For another point of view, see [S].

2.2.2.   Quantum resurgence

We consider now x as a "large" complex parameter. The formal integral,

$$\dot{\Phi}(x,z) = v_+(x)(\frac{\partial}{\partial z}q(z))^{\frac{1}{2}}\varphi_+(z,x)e^{+\frac{xz}{2}} + v_-(x)(\frac{\partial}{\partial z}q(z))^{\frac{1}{2}}\varphi_-(z,x)e^{-\frac{xz}{2}},$$

with $\varphi_\pm$ such that

$$(\frac{\partial^2}{\partial z^2} \pm x\frac{\partial}{\partial z} - H^2 + \frac{\partial}{\partial z}H)\varphi_\pm = 0,$$

can be interpreted now as a sum of two elementary symbols in x with supports respectively $\frac{z}{2}$ and $\frac{-z}{2}$. The formal power series $\varphi_\pm(x,z)$ are viewed as elements of $\mathcal{O}_{(z)}[[x^{-1}]]$, formal expansions of the powers of x^{-1}, with analytic multivalued coefficients in the variable z.

Normalisation   The quantum resurgence of the Schrödinger equation is obviously linked to the so-called *WKB method*. When going back to the variable q, the formal integral appears as a sum of two *elementary WKB symbols*

$$\dot{\Phi}_\pm(x,q) = p^{\frac{-1}{2}}(q)\varphi_\pm(x,q)e^{\pm\frac{xz(q)}{2}}$$

where $\varphi_\pm(x,q)$ belong to $\mathcal{O}_{(q)}[[x^{-1}]]$ and $p(q) = W^{\frac{1}{2}}(q)$. These elementary WKB symbols are given up to normalisation. One often chooses (see for instance [V] or [DDP1]) this normalisation by setting for q locally in a neigbourhood of q_0

$$\dot{\Phi}_{norm\pm}(x,q) = u^{\frac{-1}{2}}(x,q)\exp(\pm x\int_{q_0}^q u(x,q')\,dq'),$$

where $u(x,q)$ is a formal expansion of the powers of x^{-2},

$$u(x,q) = p(q) + \sum_{n\geq 1}x^{-2n}u_{2n}(q),$$

a solution of the equation[3],

$$x^2(W(q) - u^2) = u^{\frac{1}{2}}\frac{\partial^2}{\partial q^2}(u^{\frac{-1}{2}}).$$

Such a normalisation is said to be *a good normalisation* at q_0. By the change of variable $(q, q_0) \to (z, 0)$, the well normalised elementary WKB symbols become

$$\dot{\Phi}_{norm\pm}(x,z) = (\frac{\partial}{\partial z}q(z))^{\frac{1}{2}}U^{\frac{-1}{2}}(x,z)\exp(\pm x\int_0^z U(x,z')\,dz')$$

---

[3]One may also see $xu(x,q)$ as the odd part of the formal power series solution of the Riccati equation (7). The term $u_{2n}(q)$ is polynomial with respect to the derivatives $W^{(k)}$, $k \leq 2n$, and p^{-1}, odd in p^{-1}.

$$= (\tfrac{\partial}{\partial z}q(z))^{\frac{1}{2}}\varphi_{norm\pm}(z,x)e^{\pm\frac{xz}{2}}$$

where $U(x,z) = u(x,q(z))\tfrac{\partial}{\partial z}q(z)$. To be more precise,

$$U(x,z) = \frac{1}{2} + \sum_{n\geq 1} x^{-2n}U_{2n}(z) = \frac{1}{2} + x^{-2}(H^2 - \frac{\partial}{\partial z}H) + O(x^{-4}),$$

and moreover[4]

$$x^2(U^2 - \frac{1}{4}) = H^2 - \frac{\partial}{\partial z}H - U^{\frac{1}{2}}\frac{\partial^2}{\partial z^2}(U^{\frac{-1}{2}}).$$

Then we have

$$\varphi_{norm\pm}(x,z) = U^{\frac{-1}{2}}(x,z)\exp(\pm x\int_0^z (U(x,z') - \frac{1}{2})\,dz') = \sqrt{2} + 0(x^{-1}).$$

Remark: The well normalised symbols are related by the equality

$$\dot{\Phi}_{norm-}(x,z) = \dot{\Phi}_{norm+}(-x,z).$$

Resurgence We want here to prove the resurgence in x of the formal integral. Coming back to the formal expression (14) of $\varphi_\pm$ as an expansion of resurgence monomials $\mathcal{W}^\bullet$,

$$\mathcal{W}^\emptyset + \mathcal{W}^{(x)}(x,z) + \mathcal{W}^{(x,-x)}(x,z) + \dots$$
$$+\mathcal{W}^{(x,-x,\dots,x)}(x,z) + \mathcal{W}^{(x,-x,\dots,x,-x)}(x,z) + \dots$$

and respectively,

$$\mathcal{W}^\emptyset + \mathcal{W}^{(-x)}(x,z) + \mathcal{W}^{(-x,x)}(x,z) + \dots$$
$$+\mathcal{W}^{\overbrace{(-x,x,\dots,-x)}^{2p-1}}(x,z) + \mathcal{W}^{\overbrace{(-x,x,\dots,-x,x)}^{2p}}(x,z) + \dots.$$

We now require $\mathcal{W}^{(x,-x,\dots,\epsilon x)}$ to be *small resurgent functions* in x. Denote by $\mathcal{W}^{\overbrace{(+,-,\dots,\epsilon)}^{n}}(\xi,z)$ the minor of $\mathcal{W}^{\overbrace{(x,-x,\dots,\epsilon x)}^{n}}(x,z)$. According to (6)

---

[4]One gets also $x(U - \frac{1}{2})(x,z)$ as the odd part of the formal power series solution of the Riccati equation,
$$(\frac{\partial}{\partial z} + x)\psi = H^2 - \frac{\partial}{\partial z}H - \psi^2.$$

(see 2.1.2), we deduce from (13)

$$
\begin{cases}
(\frac{\partial}{\partial z} + \delta(\xi) + \frac{\partial}{\partial \xi})\mathcal{W}^{(+)}(\xi,z) = -H(z)\delta(\xi) <, \\
\frac{\partial}{\partial z}\mathcal{W}^{(+,-)}(\xi,z) = -H(z)\mathcal{W}^{(+)}(\xi,z) , \\
(\frac{\partial}{\partial z} + \delta(\xi) + \frac{\partial}{\partial \xi})\mathcal{W}^{(+,-,+)}(\xi,z) = -H(z)\mathcal{W}^{(+,-)}(\xi,z) , \\
\qquad \cdots
\end{cases}
$$

and in the same way,

$$
\begin{cases}
(\frac{\partial}{\partial z} - \delta(\xi) - \frac{\partial}{\partial \xi})\mathcal{W}^{(-)}(\xi,z) = -H(z)\delta(\xi), \\
\frac{\partial}{\partial z}\mathcal{W}^{(-,+)}(\xi,z) = -H(z)\mathcal{W}^{(-)}(\xi,z), \\
(\frac{\partial}{\partial z} - \delta(\xi) - \frac{\partial}{\partial \xi})\mathcal{W}^{(-,+,-)}(\xi,z) = -H(z)\mathcal{W}^{(-,+)}(\xi,z), \\
\qquad \cdots
\end{cases}
$$

Let us consider z_0 as a fixed point on the Riemann surface of H. It appears to be more convenient, in order to solve the previous equations, to impose the condition

$$(B\varphi_\pm)|_{z=z_0\pm\xi} = \delta(\xi) \pm H(z_0)$$

which allows us to fix the constants of integration. We get easily by the same method as in 2.1.2, for (ξ,z) in a neighbourhood of $(0,z_0)$,

$$
\begin{cases}
\mathcal{W}^{(+)}(\xi,z) = -H(z-\xi), \\
\mathcal{W}^{(+,-)}(\xi,z) = \int_{z_0}^{z-\xi} H(z_1)H(z_1+\xi)\,dz_1, \\
\mathcal{W}^{(+,-,+)}(\xi,z) = -\int_0^\xi \int_{z_0}^{z-\xi} H(z_1)H(z_1+\xi_1)H(z-\xi+\xi_1)\,dz_1 d\xi_1,
\end{cases}
$$

and more generally,

$$
\begin{cases}
\mathcal{W}^{(+,-,\cdots,-,+)}(\xi,z) = -\int_0^\xi \mathcal{W}^{(+,-,\cdots,-)}(z-\xi+\xi_1,\xi_1)H(z-\xi+\xi_1)\,d\xi_1, \\
\mathcal{W}^{(+,-,\cdots,-,+,-)}(\xi,z) = -\int_{z_0}^{z-\xi} \mathcal{W}^{(+,-,\cdots,-,+)}(z_1+\xi,\xi)H(z_1+\xi)\,dz_1.
\end{cases}
$$

We get also

$$
\begin{cases}
\mathcal{W}^{(-)}(\xi,z) = H(z+\xi), \\
\mathcal{W}^{(-,+)}(\xi,z) = -\int_{z_0}^{z+\xi} H(z_1)H(z_1-\xi)\,dz_1, \\
\mathcal{W}^{(-,+,-)}(\xi,z) = -\int_0^\xi \int_{z_0}^{z+\xi} H(z_1)H(z_1-\xi_1)H(z+\xi-\xi_1)\,dz_1 d\xi_1, \\
\qquad \cdots \\
\mathcal{W}^{(-,+,\cdots,+,-)}(\xi,z) = \int_0^\xi \mathcal{W}^{(-,+,\cdots,+)}(z+\xi-\xi_1,\xi_1)H(z+\xi-\xi_1)\,d\xi_1, \\
\mathcal{W}^{(-,+,\cdots,+,-,+)}(\xi,z) = -\int_{z_0}^{z+\xi} \mathcal{W}^{(-,+,\cdots,+,-)}(z_1-\xi,\xi)H(z_1-\xi)\,dz_1.
\end{cases}
$$

Remarks:

• Denote by $S_{(-1)}$ the operator defined by

$$S_{(-1)}\varphi(x) = \varphi(-x).$$

Then, if $\varphi(x) = a_0 + {}^b\varphi(x)$,

$$B(S_{(-1)}\varphi)(\xi) = a_0\delta(\xi) - \varphi(-\xi).$$

We deduce that the initial datum

$$B\varphi_{+|z=z_0+\xi} = \delta(\xi) - H(z_0),$$

becomes

$$B(S_{(-1)}\varphi_+)_{|z=z_0-\xi} = \delta(\xi) + H(z_0).$$

This allows us to assert that $\varphi_-(x, z) = \varphi_+(-x, z)$.

• Our initial datum, $(B\varphi_\pm)_{|z=z_0\pm\xi} = \delta(\xi) \pm H(z_0)$, is not the right one to get the Borel transform of the well normalised expansions $\varphi_{norm\pm}(x, z)$. Our construction gives in fact some $\varphi_\pm(x, z)$ which can be written as

$$\varphi_\pm(x, z) = a(\pm x)U^{\frac{-1}{2}}(x, z)\exp(\pm x\int_0^z(U(x, z') - \frac{1}{2})\,dz'),$$

with $a(x) = \frac{1}{\sqrt{2}} + O(x^{-1})$.

It is then easy to deduce the resurgence of the well normalised WKB symbols[5].

We refer the reader to [D2] for the details of the proof of the endless continuability of the well normalised WKB symbols. One can prove in fact that the sum of the expansion of these resurgence monomials is *endlessly continuable with respect to the two variables ξ and z* – for a convenient generalization to several variables of the notion of "endlessly continuable function" – provided that the function H is an *endlessly continuable function*. In particular, it is true when $W(q)$ is a polynomial, as in our case. Roughly speaking, the result can be summarised as the following, ([DDP1], [E2]):

Proposition 2.8 *The elementary well normalised WKB symbols,*

$$(\frac{\partial}{\partial z}q(z))^{\frac{1}{2}}\varphi_\pm(z, x)e^{\frac{\pm zx}{2}},$$

---

[5]For example,

$$U = \frac{1}{2} + \frac{1}{2x\varphi_+\varphi_-}Wronsk(\varphi_+, \varphi_-),$$

where *Wronsk* is the wronskian. Then $U(x, z)$ is resurgent ... .

are resurgent (summable) symbols. The resurgent functions $\varphi_\pm(z,x)$ *satisfy the following resurgent equations:*

$$\begin{cases} \Delta_{\overline{\omega}_+}\varphi_+ = V_{\overline{\omega}_+}(x)\varphi_- \,, \\ \Delta_{\overline{\omega}_-}\varphi_- = V_{\overline{\omega}_-}(x)\varphi_+ \,, \end{cases}$$

where $\overline{\omega}_+$ *(resp.* $\overline{\omega}_-$*) are "movable singularities"* $(z - \omega_i)$ *(resp.* $(\omega_i - z)$*) or "fixed singularities"* $\pm(z_0 - \omega_i)$*, while the* ω_i *belong to a finite set of complex numbers linked to the set of singularities of the function* H.

The functions $V_{\overline{\omega}_\pm}(x)$ are the so-called *Voros coefficients*. They are resurgent functions in x with noteworthy resurgent properties which are too long to be presented here.

We send the reader to the references for a complete study of the quantum resurgence of the Schrödinger equation. It allows us to make exact the usual semi-classical methods, thus giving the mathematical frame to the "Voros program" [V].

References

[1]   For the first section

[Ah] Ahmedou Ould Jidoumou, *Modèles de résurgence paramétrique (fonctions d'Airy et cylindro-paraboliques)*. Thèse de Doctorat, Nice 1990.

[C]   B.Candelpergher, *Une introduction à la résurgence*. Gazette des Mathématiciens (Soc. math. France), oct. 89 n°42.

[CNP] B.Candelpergher, C.Nosmas, F.Pham, *Approche de la résurgence*. Hermann 1992.

[E1] J.Ecalle, *Les fonctions résurgentes*. Publ. Math. Université Paris-Sud (3 volumes).

[E2] J.Ecalle, *Cinq applications des fonctions résurgentes*. Preprint 84T 62, Orsay.

[E3] J.Ecalle, *Finitude des cycles limites et accéléro-sommation de l'application de retour*. Bifurcations of planar vector fields (J.P.Françoise & J.C.Roussarie ed.), Lecture Notes in Maths n°1455, Springer 1990.

[E4] J.Ecalle, *Fonctions analysables et preuve constructive de la conjecture de Dulac*. Actualités mathématiques (Hermann).

[E5] J.Ecalle, *Singularités non abordables par la géométrie. Annales de l'Institut Fourier, Tome 42 (1992)-Fascicules 1 and 2.*

[Ma1] B.Malgrange, *Travaux d'Ecalle et de Martinet-Ramis sur les systèmes dynamiques. Sém. Bourbaki 1981-82, exp.n°582, Astérisque 92-93, 1982.*

[Ma2] B.Malgrange, *Introduction aux travaux de J.Ecalle. L'Enseignement Mathématique 31, pp. 261-282, 1985.*

[MR1] J.Martinet, J.P.Ramis, *Problèmes de modules pour les équations diffrentielles non linéaires du premier ordre. Publ.Math.IHES 55 (1982), pp.117-214.*

[MR2] J.Martinet, J.P.Ramis, *Théorie de Galois différentielle et resommation. Computer Algebra and Differential Equations (E.Tournier ed.), Acad.Press 1988.*

[MR3] J.Martinet, J.P.Ramis, *Théorie de Cauchy sauvage. To appear.*

[Ph1] F.Pham, *Résurgence d'un thème de Huygens-Fresnel. Publ.Math.IHES 68 (Vol. en l'honneur de R.Thom), 1988.*

[Ph2] F.Pham, *Fonctions résurgentes implicites. C. R. Acad. Sci. Paris t.309, Série I, pp. 999 - 1001, 1989.*

[2]   For the second section

[B] F.Blais, *Fleuves critiques. C. R. Acad. Sci. Paris, t. 307, Série I, pp.439-442, 1988.*

[CNP] B.Candelpergher, C.Nosmas, F.Pham, *Approche de la résurgence. Hermann 1992, and Premiers pas en calcul étranger. To appear in Annales de l'Institut Fourier.*

[D1] E.Delabaere, *Etude de l'opérateur de Schrödinger stationnaire unidimensionel pour un potentiel polynôme trigonométrique. C. R. Acad. Sci. Paris, t. 314, Série I, pp.807-810, 1992.*

[D2] E.Delabaere, *Résurgence équationnelle et résurgence quantique de l'équation de Schrödinger. To appear.*

[DD] E.Delabaere, H.Dillinger, *Contribution à la Résurgence quantique. Thèse de doctorat. Université de Nice-Sophia-Antipolis, 1991.*

[DDP0] E.Delabaere, H.Dillinger, F.Pham, *Développements semi-classiques exacts des niveaux d'énergie d'un oscillateur à une dimension. C. R. Acad. Sci. Paris, t. 310, Série I, pp.141-146, 1990.*

[DDP1] E.Delabaere, H.Dillinger, F.Pham, *Résurgence de Voros et périodes des courbes hyperelliptiques.* To appear in *Annales de l'Institut Fourier.*

[DDP2] E.Delabaere, H.Dillinger, F.Pham, *Exact semi-classical expansions for a one dimensional oscillator.* To appear.

[Di] F. & M.Diener, *Fleuves. Univ. Paris VII, 1987.*

[E1] J.Ecalle, *Les fonctions résurgentes. Publ. Math. Université Paris-Sud (3 volumes).*

[E2] J.Ecalle, *Cinq applications des fonctions résurgentes.* Preprint 84T 62, Orsay.

[F] F.Fauvet, *Etude résurgente de quelques bifurcations. Thèse de doctorat.* Université Louis Pasteur (Strasbourg I), 1992.

[Ph3] F.Pham, *Resurgence, quantized canonical transformations, and multi-instanton expansions. Algebraic Analysis (papers dedicated to M.Sato),* Vol. II, Acad.Press 1988.

[S] Y.Sibuya, *Global theory of a second order linear ordinary differential equation with a polynomial coefficient. North-Holland, Mathematics Studies 18.*

[V] A.Voros, *The return of the quartic oscillator,* Ann. Inst. H.Poincaré 29,3- 1983-.

Perturbation Analysis of Nonlinear Systems

Kenneth R. Meyer [*]

Contents

1. Introduction   104

2. Examples of equations in normal form   105

3. Reducing an equation to normal form   108

4. The general algorithm   112

5. The proof of general Lie transform algorithm   117

6. Function applications   124

7. Autonomous differential equations   126

8. Non-autonomous differential equations   129

9. Classical Hamiltonian systems   131

10. The computational Poincare Lemma and Darboux Theorem 133

11. Poisson systems   135

[*]Institute for Dynamics, Department of Mathematics, University of Cincinnati, Cincinnati, Ohio 45221-0025 —— This research partially supported by grants from the National Science Foundation.

1. Introduction

This paper is an introduction to perturbation analysis of differential equations and other nonlinear systems using normal form theory. Since it is an introduction a large part of the material presented will be classical, but the last section does contain some new applications. I hope that by understanding the classical results presented here the reader will be able to gain an entry into this rapidly evolving field. Since I am interested in applications of the theory to specific examples I will emphasize the development of the main computer algorithm and use of algebraic processors. This computer algebra approach is reflected in the presentation of the theory from a Lie transform approach. It is truly an algorithm in the sense of modern computer science: a clearly defined iterative procedure.

In this paper I would like to indicate the great genervality of the method by illustrating how it can be used to solve perturbation problems that are typically solved by other methods, often special ad hoc methods. In most cases I have chosen the simplest standard examples. In fact most of the paper consists of examples of problems that can be solved by Lie transforms, without spending too much time on the derivation or the theory. There are many topics of current research that are not considered here since this is to be an introduction not a summary of new results.

The method of Lie transforms was first given in Deprit(1969) for Hamiltonian systems of differential equations, then generalized to arbitrary systems of differential equations by Kamel(1970) and Henrard(1970). An earlier survey of the method is Meyer(1991) and a much more complete development for Hamiltonian systems only can be found in Meyer and Hall(1992). Also see Chow and Hale(1982) and Golubitsky and Schaeffer (1985).

In sections 2 and 3 an overview of the field is given by considering the classical special case of normal forms for an autonomous differential equation with a semi-simple linear part. There are many interesting results for this special case. In Section 2 it is assumed that the equations are actually in normal form whereas in Section 3 it is assumed that the equations are in normal form only to a certain order in ϵ.

Section 4 presents the general method of Lie transforms as it applies to general tensor fields be they functions, vector fields or Poisson structures. There are no proofs give here. Section 5 contains the technical proofs of the perturbation theorems of Lie transform theory.

The remaining sections give further examples of applications of the method to various tensor fields. Section 6 gives applications to functions – a computational implicit function theorem and splitting lemma. Section 7 gives further applications of the theory to differential equations including the General Equilibrium Theorem. Section 8 contains applications to non-autonomous differential equation – the method of averaging and a computational Liapunov transformation. Section 9 gives an application to Hamiltonian systems. Section 10 gives a computational Poincaré lemma and Darboux theorem. And Section 11 gives a new application to Poisson systems.

2.   Examples of equations in normal form

Consider the equation

$$\dot{\xi} = A\xi + \epsilon f(\xi, \epsilon) \tag{1}$$

where $\xi \in \mathbf{R}^m$, A is an $m \times m$ constant matrix, f is analytic in ξ and ϵ for ξ in some domain and ϵ small. Here and below the independent variable is t, time, and the dot indicates differentiation with respect to t. The variable ϵ is considered as a small parameter. Low differentiability questions will be ignored in this introduction. Assume that A is diagonalizable – sometimes A is said to be semi-simple or just simple. Then classically (1) is said to be in *normal form* if f satisfies

$$f(e^{At}\xi, \epsilon) \equiv e^{At} f(\xi, \epsilon). \tag{2}$$

The simplest normal form would be $f \equiv 0$ so that the equation would be linear and thus solvable. As the next simplest case consider the problem of a perturbation of a harmonic oscillator. If $A = \begin{pmatrix} 0 & 1 \\ -1 & 0 \end{pmatrix}$ and $\xi = (u,v)^T$ then the equation is in normal form if the equation is of the form

$$\dot{u} = v + \epsilon\{+u\alpha(u^2 + v^2, \epsilon) + v\beta(u^2 + v^2, \epsilon)\},$$

$$\dot{v} = -u + \epsilon\{-u\beta(u^2 + v^2, \epsilon) + v\alpha(u^2 + v^2, \epsilon)\}, \tag{3}$$

where $\alpha(\zeta, \epsilon)$ and $\beta(\zeta, \epsilon)$ are analytic in the two variables ζ and ϵ. Note that when $\epsilon = 0$, (3) is the system corresponding to the equation $\ddot{u} + u = 0$, the harmonic oscillator. These equations are not easy to understand in rectangular coordinates but they are in polar coordinates or complex variables. In polar coordinates equations (3) become

$$\dot{r} = \epsilon r\alpha(r^2),$$

$$\dot{\theta} = -1 - \epsilon\beta(r^2), \tag{4}$$

and in complex coordinates $w = re^{-i\theta}, \bar{w} = re^{i\theta}$ the equations become

$$\dot{w} = iw + \epsilon w\gamma(w\bar{w}),$$

$$\dot{\bar{w}} = -i\bar{w} + \epsilon\bar{w}\bar{\gamma}(w\bar{w}), \tag{5}$$

where $\gamma(\xi) = \alpha(\xi) + i\beta(\xi)$. In (5) the second equation is just the conjugate of the first when α and β are real.

Polar coordinates are the easiest to use in this example. Let the subscript zero denote the initial condition at time $t = 0$. When $\epsilon = 0$ the solution of (4) lies on a circle $r = r_0 = $ constant and the angle θ decreases uniformly with t; i.e. $\theta(t) = \theta_0 - t$. All solutions are periodic with constant period 2π. A solution moves on a circle in the (u, v)-plane with uniform angular frequency equal to 1. This is the phase portrait of a linear center.

Now consider the case when $\alpha \equiv 0$, but $\epsilon \neq 0$. A solution of (4) is $r = r_0 = $ constant, $\theta(t) = \theta_0 - t(1 + \epsilon\beta(r_0))$. This solution still lies on a circle and θ decreases uniformly with t, but now the angular frequency depends on the circle. All solutions are periodic but the period changes with the amplitude. This is the general center. The function β gives the dependence of the frequency on the amplitude.

If $\alpha \not\equiv 0$ but $\alpha(r') = 0$ for some specific value of $r = r'$ then $r(t) \equiv r', \theta(t) = \theta_0 - t(1 + \epsilon\beta(r'))$ is a specific periodic solution with amplitude r' and period $2\pi/(1 + \epsilon\beta(r'))$. If $\alpha(r') = 0$, $\alpha(r'') = 0$ and $a(r) > 0$ when $r \in (r', r'')$ then a solution $(r(t), \theta(t))$ that starts with $r_0 \in (r', r'')$ will have $r(t) \to r''$ as $t \to +\infty$ and $r(t) \to r'$ as $t \to -\infty$. The periodic solution where $r = r'$ is a limit cycle in the plane. The frequency is controlled by α and the amplitide by β.

Consider the general equation (1). Consider the difficult perturbation problem when the linear system is many harmonic oscillators with the same period T, so $\exp(AT) = I$. With f satisfying (2) it is easy to find periodic solutions. If $\xi = \zeta(\epsilon)$ satisfies the equations

$$\tau A\xi = f(\xi, \epsilon) \tag{6}$$

then $e^{A(1+\epsilon\tau)t}\zeta(\epsilon)$ is a $T/(1 + \epsilon\tau)$ periodic solution of (1). Thus finding a periodic solution is reduced to solving the equations (6).

Now consider the Hamiltonian case. Let equation (1) be Hamiltonian, i.e. $\xi \in \mathbf{R}^{2n}$ and there is a function

$$H(\xi, \epsilon) = \frac{1}{2}\xi^T S\xi + \epsilon F(\xi, \epsilon) \tag{7}$$

such that equation (1) is

$$\dot{\xi} = J\nabla H(\xi, \epsilon) = JS\xi + \epsilon J\nabla F(\xi, \epsilon) \tag{8}$$

where J is the standard $2n \times 2n$ skew symmetric matrix

$$J = \begin{pmatrix} 0 & I \\ -I & 0 \end{pmatrix}. \tag{9}$$

Again assume that $A = JS$ is simple, i.e. diagonalizable. Then the Hamiltonian is in normal form if $H(\xi, \epsilon)$ is constant on the trajectories of the linear flow, i.e.

$$H(e^{At}\xi, \epsilon) \equiv H(\xi, \epsilon) \text{ or } F(e^{At}\xi, \epsilon) \equiv F(\xi, \epsilon). \tag{10}$$

This is equivalent to the condition (2) for the equations of motion (8).

Consider the case where the unperturbed system, when $\epsilon = 0$, is two harmonic oscillators with distinct frequencies ω_1 and ω_2. Then without loss of generality

$$H(\xi, 0) = \frac{1}{2}\xi^T S\xi = (\omega_1/2)(u_1^2 + v_1^2) + (\omega_2/2)(u_2^2 + v_2^2), \tag{11}$$

where $\xi = (u_1, u_2, v_1, v_2)$. If the Hamiltonian is not of this form then a linear symplectic change of variables brings it to this form, see Birkhoff(1926) or Meyer and Hall(1992). You cannot assume that both ω_1 and ω_2 are positive. The Hamiltonian and its flow are easier to understand if action-angle variables are used. They are just the Hamiltonian or symplectic polar coordinates. Introduce action-angle variables, $(I_1, I_2, \phi_1, \phi_2)$, by

$$I_i = (1/2)(u_i^2 + v_i^2), \qquad \phi_i = \arctan(y_i/x_i). \tag{12}$$

In these coordinates the condition (10) is just

$$H(I_1, I_2, \phi_1 + \omega_1 t, \phi_2 + \omega_2 t, \epsilon) \equiv H(I_1, I_2, \phi_1, \phi_2, \epsilon) \tag{13}$$

and if ω_1/ω_2 is irrational (13) implies that H is independent of the angles ϕ_1, ϕ_2. Consider this case first. In this case the equations of motion are

$$\dot{I}_1 = \frac{\partial H}{\partial \phi}_1 = 0, \quad \dot{\phi}_1 = -\frac{\partial H}{\partial I}_1 = -\omega_1 + \epsilon \alpha_1(I_1, I_2, \epsilon),$$

$$\dot{I}_2 = \frac{\partial H}{\partial \phi}_2 = 0, \quad \dot{\phi}_2 = -\frac{\partial H}{\partial I}_2 = -\omega_2 + \epsilon \alpha_2(I_1, I_2, \epsilon), \tag{14}$$

where $\alpha_i = \partial F/\partial I_i$. The two action variables I_1, I_2 are integrals for the flow. The invariant set $I_1 = 0$ is the same as $u_1 = v_1 = 0$ and so represents the (u_2, v_2)-plane in $\mathbf{R}^4$. In that plane the flow is the same as the general center discussed above. Similarly for $I_2 = 0$.

The invariant set $I_1 = I_{10} > 0$ and $I_2 = I_{20} > 0$ is a torus in $\mathbf{R}^4$ and the angles ϕ_1 and ϕ_2 are coordinates on this torus. The flow on this torus is the linear flow given by

$$\phi_1(t) = \phi_{10} - (\omega_1 - \epsilon\alpha_1(I_{10}, I_{20}, \epsilon))t,$$

$$\phi_2(t) = \phi_{20} - (\omega_2 - \epsilon\alpha_2(I_{10}, I_{20}, \epsilon))t. \tag{15}$$

If the slope of this line in the (ϕ_1, ϕ_2)-plane is rational then the solutions are periodic and are closed curves on the torus. If the slope is irrational then the solutions are quasi-periodic and are dense curves on the torus. Of course in general the slope changes from torus to torus. Thus all of $\mathbf{R}^4$ is made up of two invariant planes filled with periodic solutions and invariant tori with either periodic or quasi-periodic solutions.

When ω_1/ω_2 is rational then the normalized Hamiltonian contains some angle terms. These terms are very important in understanding bifurcations of periodic solutions. There are many fascinating bifurcations which occur when ω_1/ω_2 is a rational number. See Meyer and Schmidt(1971) for the case when $\omega_1/\omega_2 = 1$, the so-called Hamiltonian-Hopf Theorem, and Schmidt(1974) for the cases where ω_1/ω_2 is a rational other than 1.

3.   Reducing an equation to normal form

In the last section several simple examples of equations in normal form were given and they were easy to analyze, but unfortunately the equations that come from real problems in mathematics, physics or engineering are seldom in normal form. So the problem is how to reduce an equation to normal form and how much information can be gleaned from the normal form. This section contains a few such results.

Sometimes a problem comes with a parameter that can be considered small at least in some cases and sometimes a small parameter must be introduced into the problem. Van der Pol's equation is

$$\ddot{u} + \epsilon(u^2 - 1)\dot{u} + u = 0 \tag{1}$$

and written as a system it is

$$\dot{u} = v,$$

$$\dot{v} = -u + \epsilon(1 - u^2)v. \tag{2}$$

The original physical system modeled by this equation was a vacuum tube tuning circuit for a radio. See van der Pol(1926). The parameter was a dimensionless ratio which may or may not be small. In this case the problem comes with a parameter.

Consider an autonomous differential equation which has an equilibrium point at the origin. The equation can be written in the form

$$\dot{x} = Ax + g(x) \tag{3}$$

where g is the quadratic and higher order terms in the Taylor expansion of the equation about the origin, i.e. $g(0) = 0$ and $Dg(0) = 0$ where Dg is the Jacobian matrix of g. In order to study the solutions of (3) near the equilibrium point scale the variable x by $x \to \epsilon x$, that is, make the change of variables $x = \epsilon x'$ in the equation and then drop the prime. The equation (3) becomes

$$\dot{x} = Ax + \epsilon h(x, \epsilon) \tag{4}$$

where $h(x, \epsilon) = g(\epsilon x)/\epsilon^2$. Since g contains quadratic and higher degree terms the function h will be analytic in ϵ.

Henceforth when referring to equation (4) it will not be assumed that it came from an equation of the form (3) but that the h has a series expansion in ϵ whose coefficients are polynomials in x. Thus van der Pol's equation (2) will be considered of the form (4).

Exercise: Take the pendulum equation $\ddot{\theta} + \sin\theta = 0$, write it as a system, note that it has a critical point at the origin, write it in the form (4). It also has an equilibrium point at $\theta = n\pi, \dot{\theta} = 0$. Shift this point to the origin and write the equation in the form (4).

Several systems can be reduced to the form (4), but in general the system is not in normal form. One approach is to try to construct a change of variables which reduces the equation to normal form. There is a vast literature on this problem which is made of many partial solutions to the question of the existence of a normalizing transformation. The first and simplest result is the formal theorem.

Classical Normalization Theorem. *Assume that A is diagonalizable. Then there is a formal change of variables $x = \xi + \cdots$ which is a formal series in ϵ which reduces (3.4) to normal form (2.1) with (2.2) holding.*

In general the series for the change of variable diverges, but there are a few cases where the series actually converges.

Poincare's Linearization Theorem. *Let the system (3) be analytic, A diagonalizable with eigenvalues $\lambda_1, ..., \lambda_m$. Assume that the eigenvalues of A have negative real parts and that there is no relation between the eigenvalues of the form $\lambda_j = k_1\lambda_i + \cdots + k_m\lambda_m$ where $k_i \in Z, k_1 + \cdots + k_m \geq 2$. Then there is an analytic change of variables $x = \xi + \cdots$ which linearizes the equation, i.e. equation (3) in the new coordinates ξ is just the linear equation $\dot{\xi} = A\xi$.*

The condition $\lambda_j \neq k_1\lambda_i + \cdots + k_m\lambda_m$ is necessary for the existence of a formal linearizing transformation – see Section 7. In fact the system

$$\dot{u} = u,$$

$$\dot{v} = 2v + \alpha u^2$$

$$(5)$$

with $\alpha = 1$ cannot be reduced to a linear system by a C^2 transformation.

Exercise. Note that the example (2) is in normal form and is solvable by elementary means. Solve the equation with $\alpha = 0$ and with $\alpha = 1$. Note the difference in the differentiability of the solution curves at the origin. Show that there is not a C^2 linearizing transformation. See Sternberg (1959). Hartman(1964) gives an example of a system that is also in normal form but is not C^1 linearizable. (Many counter examples are equations in normal form – Meyer(1986).)

There are many variations of this theorem. Siegel proved an analytic linearization theorem with the two conditions on the eigenvalues replaced by an inequality of the form $|\ k_1\lambda_i + \cdots + k_m\lambda_m\ | > K/\ |\ k_1 + \cdots + \lambda_m\ |^\gamma$. Sternberg(1959) started a whole industry of proving C^k linearization theorems.

Siegel proved that in general the normalizing series diverges. The best results on the actual convergence and divergence of the series are in Bruno(1971,1972).

Since the normalizing series diverges in general why are people using normal forms? The answer is that one only needs to put the system in normal form up to a finite order to obtain some interesting results and the transformation to normalize is polynomial, so convergent. In fact the real theorem of interest is the following.

Classical Finite Normalization Theorem. *Assume that A is diagonalizable. Let N be a given positive integer. There is a change of variables* $x = \xi + \cdots + \epsilon^N c(x)$ *which is polynomial in ϵ of degree N with coefficients which are polynomials in x which reduces (3.4) to normal form up to order N, i.e. which reduces (3.4) to*

$$\dot{\xi} = A\xi + \epsilon f(\xi, \epsilon) + \epsilon^{N+1} k(\xi, \epsilon)$$

$$(6)$$

where f satisfies

$$f(e^{At}\xi, \epsilon) \equiv e^{At} f(\xi, \epsilon).$$

$$(7)$$

Since the original equation and the transformation are analytic so are f and k. The *truncated system* is obtained from (6) by forgetting the non-normalized terms, i.e. the truncated system is

$$\dot{\xi} = A\xi + \epsilon f(\xi, \epsilon).$$

$$(8)$$

Since this system is in normal form it is easy to analyze and the standard continuity theorem for ordinary differential equations ensures that the solutions of (6) and (8) are within terms of the order $O(\epsilon^{N+1})$ for bounded times.

My favorite theorem in this subject is Poincaré's Center Theorem. Consider first an interesting case not covered by Poincaré's Linearization Theorem, i.e. consider the case where $m = 2$ and $A = \begin{pmatrix} 0 & 1 \\ -1 & 0 \end{pmatrix}$. This is the perturbation of the harmonic oscillator. Since the eigenvalues of A are $\pm i$ there are infinitely many relations between the eigenvalues of the form $i = (k+1)i + k(-i)$. The normal form is as discussed in the previous section – see (2.3, 2.4, 2.5). Think about the polar form (2.4). The function α determines the stability of the origin. If after a finite number of steps a non-zero term in the expansion of α is found then it is easy to determine the sign of the derivative of r near the origin even in the presence of the non-normalized terms. If $\dot{r} > 0$ then the origin is asymptotically unstable and if $\dot{r} < 0$ then the origin is asymptotically stable. So the question remains as to the stability of the origin if formally α is identically zero. Poincaré(1885) proved in this case that the normalizing series actually converges so the system has a center at the origin! This theorem is known as Poincaré's Center Theorem.

There has been some controversy about this theorem. Poincaré assumes that the original system is simply polynomial and some people have had trouble reading his proof. However, at this conference Y. Sibuya and R. Schafke assure me that they carefully went through the original proof and that they were able to verify that the proof is correct and in fact works for analytic systems as well as polynomial systems.

Another example is van der Pol's equation (1) or (2). In normal form up to first order, $N = 1$, it is

$$\dot{r} = \epsilon(r/2)(1 - r^2/4) + O(\epsilon^2),$$
$$\dot{\theta} = -1 + O(\epsilon^2). \tag{9}$$

The truncated system is the same as one of the examples from the previous section.

In order to find periodic solutions first consider the truncated equations. Say that a periodic solution of the truncated system has been found. One characteristic multiplier of a periodic solution of an autonomous equation is always +1, if the others differ from +1 before ϵ^{N+1} then the solution can be continued into the full system. Specifically:

Henrard's Theorem. *Let $\phi(t, \epsilon)$ be a family of periodic solutions of the truncated equation (8) of period $T(\epsilon)$. Let the characteristic multipliers be*

$\lambda_1 = +1, \lambda_2, \ldots, \lambda_m$. *If $\lambda_i - 1$ is not $O(\epsilon^{N+1})$ then the full equation (6) has a family of periodic solutions $\psi(t, \epsilon) = \phi(t, \epsilon) + O(\epsilon^{N+1})$ of period $T(\epsilon) + O(\epsilon^{N+1})$.*

Exercise. Apply this result to van der Pol's equation.

Probably the most famous results which depend on normal forms are the KAM theorems. As an example consider the problem of determining the stability of the origin for a Hamiltonian system whose linearization is two harmonic oscillators. That is, consider a system of the form

$$H(I_1, I_2, \phi_1, \phi_2) = \omega_1 I_1 + \omega_2 I_2 + \cdots \tag{10}$$

where I_1, I_2, ϕ_1, ϕ_2 are action-angle variables as discussed in the previous section. If ω_1 and ω_2 are both positive then the Hamiltonian is positive definite and so by a simple Liapunov argument the origin is stable. However, if the ω's are of different sign then the stability depends on the higher order terms. This occurs in the restricted three body problem at the Lagrange equilateral triangular points for example.

Cherry(1928) gave an example of a polynomial system in normal form that is of the above form with $\omega_1/\omega_2 = -1/2$ but for which the origin was unstable. However, if H has been normalized through the fourth order terms so that

$$H(I_1, I_2, \phi_1, \phi_2) = \omega_1 I_1 + \omega_2 I_2 + (1/2)(AI_1^2 + 2BI_1 I_2 + CI_2^2) + \cdots \tag{11}$$

then Arnold's Theorem can be applied.

Arnold's Theorem. *Let H be as above with $D = A\omega_2^2 - 2B\omega_1\omega_2 + C\omega_1^2$ nonzero. Then the origin is stable. In fact the origin is enclosed in invariant tori in the energy level $H = 0$.*

In general to be able to normalize the system to the form (11) it must be assumed that $|\omega_1/\omega_2| \neq 1, 1/2, 1/3, 1/4$. In this theorem the fabled small divisors are overcome. After over thirty years of work even the simple proofs are obscure. See Siegel and Moser(1971) and Herman(1979,1983,1986).

4.   The general algorithm

This section contains an introduction to the general algorithm and the next section contains the technical proofs.

As has been shown in the previous sections the traditional setting of perturbation theory is a differential equation depending on a small parameter ϵ. When $\epsilon = 0$ the differential equation is simple and well understood, say for example a harmonic oscillator. The problem is to understand the solutions of the equations when ϵ is nonzero but small. But there are objects different from differential equations to which perturbation analysis can be applied.

To gain generality think of any smooth tensor field defined on some open set $D \subset \mathbf{R}^m$ depending on a small parameter. The tensor field might be a function; a contravariant vector field, i.e. an ordinary differential equation; a covariant vector field, i.e. a differential form; a Riemannian metric; a Poisson structure; or any of the other classical tensors of differential geometry. The important thing about these objects is that there is a Lie derivative defined for them.

Let F be a smooth tensor field defined on an open set $D \in \mathbf{R}^m$, that is for each point $x \in D$ there is assigned a unique tensor, $F(x)$, of a fixed type, say p-covariant and q-contravariant. Let W be a smooth autonomous ordinary differential equation defined on D, i.e. a contravariant vector field on D, and let $\phi(\tau, \xi)$ be the solution of the equation which satisfies $\phi(0, \xi) = \xi$. The Lie derivative, $\mathcal{L}_W F$, is simply the directional derivative of F in the direction of W and is a tensor field of the same type as F itself. The general definition is given in any non-elementary book on differential geometry. Here I shall simply give examples.

Differential geometry has used many different notations which still persist today making a general presentation difficult. For example the object W given above might be called an autonomous differential equation on D and so W is thought of as a smooth function from D into $\mathbf{R}^m$ and is denoted by

$$\frac{dx}{d\tau} = W(x). \tag{1}$$

Then W is considered as a column vector with components $W_1, \ldots, W_m$. In classical tensor terminology W is 1-contravariant and we write W^i where i is a free index ranging from 1 to m – here the superscript tells you it is contravariant. More recent notation is

$$W = \sum_{i=1}^{n} W^i(x) \frac{\partial}{\partial x_i} = W^1(x) \frac{\partial}{\partial x_1} + \cdots + W^m(x) \frac{\partial}{\partial x_m}. \tag{2}$$

In any case let $\phi(\tau, \xi)$ be the solution satisfying the initial condition $\phi(0, \xi) = \xi$. The simplest tensor field is a smooth function $f : D \to \mathbf{R}^1$, i.e. to each point of D you assign a scalar. The Lie derivative of f along W, $\mathcal{L}_W f$, is a smooth function from D to $\mathbf{R}^1$ also and is defined by

$$\mathcal{L}_W f(x) = \frac{\partial}{\partial \tau} f(\phi(\tau, \xi)) \mid_{\tau=0} = \nabla f(x) \cdot W(x), \tag{3}$$

the dot product of the gradient of f and W.

The next simplest tensor field is a vector field – covariant or contravariant. First let χ be a contravariant vector field or differential equation on D. Using differential equation notation for χ we write

$$\chi: \quad \dot{x} = F(x). \tag{4}$$

The column vector F is a representation of the contravariant vector field χ in the x coordinates. Do not then confuse t and τ they are different parameters for different vector fields. Changing variables in (4) from x to ξ by $x = \phi(\tau, \xi)$ where τ is simply a parameter we get

$$\dot{\xi} = \left(\frac{\partial \phi}{\partial \xi}(\tau, \xi)\right)^{-1} F(\phi(\tau, \xi)) = G(\tau, \xi). \tag{5}$$

G is the representation of χ in the new coordinates system ξ. The Lie derivative, $\mathcal{L}_W \chi$, is defined by

$$\mathcal{L}_W \chi(x) = \frac{\partial}{\partial \tau} G(\tau, \xi) \mid_{\tau=0} = \frac{\partial F}{\partial x}(x) W(x) - \frac{\partial W}{\partial x}(x) F(x). \tag{6}$$

Note that x and ξ are the same when $\tau = 0$. $\mathcal{L}_W \chi$ is a smooth contravariant vector field on D. We usually abuse the notation and confuse the vector field χ with its representation F in a coordinate system by writing $\mathcal{L}_W F$ for (6).

Let η be a 1-covariant vector field on D, i.e. a differential form, so

$$\eta = \sum_{i=1}^{m} h_i(x) dx^i. \tag{7}$$

Think of h as the column vector $(h_1, \ldots, h_m)^T$ and change variables from x to ξ by $x = \phi(\tau, \xi)$ to get

$$\eta = \sum_{i=1}^{m} k_i(\xi) d\xi^i \tag{8}$$

where k is a column vector related to h by

$$k(\tau, \xi) = \frac{\partial \phi}{\partial \xi}(\tau, \xi)^T h(\phi(\tau, \xi)). \tag{9}$$

The vector k is the components of the differential form η in the new coordinates ξ. The Lie derivative of η in the direction of W, $\mathcal{L}_W \eta$, is a one form whose component vector is given by

$$\mathcal{L}_W \eta(x) = \frac{\partial}{\partial \tau} k(\tau, \xi) \mid_{\tau=0} = \frac{\partial h}{\partial x}(x)^T W + \frac{\partial W}{\partial x}(x)^T h(x). \tag{10}$$

The Lie derivatives of other tensor fields in the direction W are defined in the same way and the reader can find a complete discussion in a book on differential geometry.

Let $T_{pq} = T_{pq}(D)$ denote the vector space of all smooth p-covariant and q-contravariant tensor fields D. A symmetric notation for $\mathcal{L}_W K$ is $[K, W]$, the Lie bracket of K and W. For fixed W the map $\mathcal{L}_W = [\cdot, W]$ is a linear operator from T_{pq} into itself. The set, $V = V(D) = T_{01}(D)$, of all smooth contravariant vector fields on D is a vector space and $[K, \cdot]$, for fixed K, is a linear from V into T_{pq}. Thus $[\cdot, \cdot] : T_{pq} x V \to T_{pq}$ is bilinear.

Suppose that the perturbation problem is given as a tensor field Z_* on D which has a formal expansion in a small parameter ϵ. In many cases ϵ is simply a scale parameter. Consider

$$Z_* = Z_*(x, \epsilon) = \sum_{j=0}^{\infty} \left(\frac{\epsilon^j}{j!} \right) Z_j^0(x) \tag{11}$$

where each Z_j^0 is a tensor field of fixed type. Specifically assume that

$$Z_j^0 \in P_j \subset T_{pq}, \text{ for } j = 0, 1, 2, \dots \ , \tag{12}$$

where P_j is a linear subspace of T_{pq}. In order to simplify the problem the method of normal forms seeks a near identity change of variables of the form $x = \xi + O(\epsilon)$ such that the tensor field Z_* in the new coordinates is simpler. The traditional approach is simple: assume a general series for the change of variables, substitute it in the series for Z_*, collect terms, and try to choose the coefficients in the change of variables series so that the tensor Z_* in the new coordinates is as simple as possible. For simple problems that will suffice, however there are several disadvantages to this approach. The bookkeeping of the terms of the series can become a major problem especially if the problem has some special structure or symmetry. For example if Z_* is a Hamiltonian vector field one would want the vector field in the new coordinates to be Hamiltonian also. Or if Z_* is invariant under some symmetry group one would want this to be true in the new coordinates also. Figuring out what the form of the nth term in new series can be quite difficult using the straight plug and chug method. Also, this procedure is not easily codable in a symbolic computer language.

Hori(1966) was interested in perturbation theory for Hamiltonian vector fields and suggested that the near identity transformation be given as the solution of an autonomous ordinary differential equation depending on a parameter. Unfortunately, not all near identity transformations are solutions of autonomous equations and so Hori was not able to develop a general theory. Deprit(1969) took Hori's idea one step further by using non-autonomous equations. He was able to give a simple set of recursive formulas that overcomes the objections given above. Hori and Deprit worked with Hamiltonian systems, but soon afterwards Kamel(1970) and Henrard(1970) considered the general case.

Thus to simplify the perturbation problem given by Z_* in (11) we seek a near identity change of coordinates of the form

$$x = x(\xi, \epsilon) = \xi + \cdots \qquad (13)$$

where $x(\xi, \epsilon)$ is constructed as a formal solution of the system of equations and initial conditions

$$\frac{dx}{d\epsilon} = W(x, \epsilon) = \sum_{j=0}^{\infty} \left(\frac{\epsilon^j}{j!} \right) W_{j+1}(x), \quad x(0) = \xi. \qquad (14)$$

It can easily be shown that for any change of coordinates of the form (13) there is a unique differential equation of the form (14) for which it is the solution function. The W above is a smooth vector field on D for each ϵ, so we take

$$W_j \in \mathcal{R}_i \subset \mathcal{V}, \text{ for all } i = 0, 1, 2, \ldots, \qquad (15)$$

where $\mathcal{R}_i$ is a linear subspace of $\mathcal{V}$, the space of smooth vector fields on D. The problem defined by Z_* may have some special symmetry, like a reflective symmetry, or a special structure, like being Hamiltonian, and this is reflected in the assumption that we have identified the subspace $\mathcal{P}_i$ to which the Z_i belong. To preserve this symmetry or structure it may be necessary to restrict the change of variables by requiring the W_i to lie in the subspaces $\mathcal{R}_i$.

In the new coordinates ξ the tensor $Z_*(x, \epsilon)$ becomes

$$Z^* = Z^*(\xi, \epsilon) = \sum_{j=0}^{\infty} \left(\frac{\epsilon^j}{j!} \right) Z_0^j(\xi). \qquad (16)$$

We say (13) or (14) transforms (11) into (16). The method of Lie transforms introduces a double indexed array $\{Z_j^i\}$ which agrees with the previous definitions when either i or j is zero. These quantities are related by the recursive formula

$$Z_j^i = Z_{j+1}^{i-1} + \sum_{k=0}^{j} \binom{j}{k} [Z_{j-k}^{i-1}, W_{k+1}]. \qquad (17)$$

The interdependence of the functions $\{Z_j^i\}$ can easily be understood by considering the Lie triangle

$$
\begin{array}{ccccc}
Z_0^0 & & & & \\
\downarrow & & & & \\
Z_1^0 & \rightarrow & Z_0^1 & & \\
\downarrow & & \downarrow & & \\
Z_2^0 & \rightarrow & Z_1^1 & \rightarrow & Z_0^2 \\
\downarrow & & \downarrow & & \downarrow
\end{array} \qquad (18)
$$

The coefficients of the expansion of the old tensor field Z_* are in the left column and those of the new tensor field Z^* are on the diagonal. The formula (17) says that to calculate any element in the Lie triangle you need the entries in the column one step to the left and up. In practice you are given Z_* and you want to find W so that Z_* is simple. The main theorem gives the definition of 'simple' as $Z_0^j \in \mathcal{Q}_j$ and so one must be able to solve the Lie equation given in (19) in the chosen spaces. Also we shall say the tensor Z^* in (16) is in *normal form* if

$$
Z_0^i \in \mathcal{Q}_i \subset \mathcal{P}_i, \text{ for } i = 1, 2, 3, \dots . \qquad (19)
$$

The fundamental theorem of the theory is:

The Normalization Theorem *Assume i)* $[\mathcal{P}_i, \mathcal{Q}_j] \subset \mathcal{P}_{i+j}$; $i, j = 1, 2, 3, \dots,$ *and ii) for any* $i = 1, 2, 3, \dots$ *and for any* $A \in \mathcal{P}_i$ *there exists* $B \in \mathcal{Q}_i$ *and* $C \in \mathcal{R}_i$ *such that*

$$
B = A + [C, Z_0^0]. \qquad (20)
$$

Then one can compute a formal expansion for W *as given in (14) with* $W_i \in \mathcal{R}_i$ *for all* i *which transforms (11) to (16) where* $Z_0^i \in \mathcal{Q}_i$ *for all* i.

In practice Z_0^0 is given and so one takes the subspaces $\mathcal{P}_i$ as small as possible. The spaces $\mathcal{Q}_i$ and $\mathcal{R}_i$ come from an analysis of the equation in (20). I will call equation (20) the *Lie equation* and the linear operator $\mathcal{L} = [\cdot, Z_0^0] : \mathcal{R}_i \rightarrow \mathcal{P}_i$ the *Lie operator*.

A development of the general Lie transform recursion formulas and the proof of this theorem are given in the next section. Since it is highly technical the reader may want to skip this section on the first reading and go to the examples.

5.   The proof of general Lie transform algorithm

In this section the main algorithm of Deprit and the main perturbation algorithm for general tensor fields are established. A general reference for the tensor analysis and notation used here is Abraham and Marsden(1978).

Let E, F, G and $E_1, \ldots, E_k$ be vector spaces over K where K is the real numbers R or the complex numbers C, $L(E; F)$ be the space of bounded linear functions from E to F, $E^* = L(E, K)$ be the dual space of E, and $L^k(E_1, \ldots, E_k; K)$ be the space of bounded multilinear maps from $E_1 \times \cdots \times E_k$ into K. Define $T_s^r(E) = L^{r+s}(E^*, \ldots, E^*, E, \ldots, E; K)$ – r copies of E^* and s copies of E, so if $Z \in T_s^r(E)$ then $Z : E^* \times \ldots \times E^* \times E \times \ldots \times E \to K$ is linear in each argument. The elements, $Z \in T_s^r(E)$, are called r-contravariant, s-covariant tensors or simply (r, s)-tensors. In the case $r = s = 0$ we define $T_0^0(E) = K$. If $A : E \to E$ is an invertible linear map and $A^* : E^* \to E^*$ is the dual map, then $A_s^r : T_s^r(E) \to T_s^r(E)$ is the invertible linear map defined by $(A_s^r Z)(\alpha^1, \ldots, \alpha^r, \beta_1, \ldots, \beta_s) = Z(A^*\alpha^1, \ldots, A^*\alpha^r, A^{-1}\beta_1, \ldots, A^{-1}\beta_s)$.

Let M be a smooth manifold modeled on a vector space E and $p \in M$ any point. In the classical and still most important case M is simply an open set D in R^m and E is R^m itself. The tangent space to M at p, denoted by T_pM, is isomorphic to E itself; the cotangent space to M at p, denoted by T_p^*M, is the dual of T_pM; and the space of r-contravariant, s-covariant tensors at p is $T_s^r(T_pM)$. The vector bundles built on T_pM, T_p^*M, and $T_s^r(T_pM)$ are respectively: TM, the tangent bundle; T^*M, the cotangent bundle; and T_s^rM, the (r, s)-tensor bundle. Smooth sections in these bundles are called respectively: vector fields (or contravariant vector fields or ordinary differential equations); covector fields (or one forms); and (r, s)-tensor fields. Let $T(M)$ be the space of smooth vector fields, $T^*(M)$ the space of smooth one forms, and $T_s^r(M)$ the space of smooth (r, s)-tensors. Let $V : M \to M$ be a diffeomorphism, $p \in M, q = V(p)$ and $DV(p) : T_pM \to T_qM$ be the derivative of V at p; then $DV_s^r(p) : T_s^r(T_pM) \to T_s^r(T_qM)$. The results of this section are quite general so M could be a Banach manifold modeled on a reflexive Banach space E, but I have no examples which require this level of generality.

Consider the case where M is an open set in R^m with coordinates $(x^1, \ldots, x^m)$. A $(0, 0)$-tensor field is simply a smooth function $Z : M \to K$. A vector field, Z, is given by

$$Z = Z^1(x)\frac{\partial}{\partial x^1} + \cdots + Z^m(x)\frac{\partial}{\partial x^m}, \tag{1}$$

where $Z^1, \ldots, Z^m$ are smooth real valued functions on M. The vector field Z is the same as the differential equation

$$\dot{x} = Z(x) \quad (\text{ or } \dot{x}^i = Z^i(x), \ i = 1, \ldots, m). \tag{2}$$

A covector field, Z, is given by

$$Z = Z_1(x)dx^1 + \cdots + Z_m(x)dx^m, \tag{3}$$

where again $Z_1, \ldots, Z_m$ are smooth functions.

Let U be a smooth vector field (autonomous differential equation) on M and let $X(\tau, y)$ be the general solution of the differential equation

$$x' = \frac{dx}{d\tau} = U(x) \tag{4}$$

which satisfies $X(0, y) = y$. That is, $X'(\tau, y) = U(X(\tau, y))$. Assume that there is an $\tau_0 > 0$ such that $X : (-\tau_0, \tau_0) \times M \to M$ is defined and smooth. X is a function of two arguments and let $'$ denote the partial derivative with respect to the first argument, $' = \partial/\partial\tau$, and let D denote the partial derivative with respect to the second argument, $D = \partial/\partial y$, thus $DX(\tau, p) : T_pM \to T_qM, q = X(\tau, p)$ and $DX_s^r(\tau, p) : T_s^r(T_pM) \to T_s^r(T_qM)$. Let $Z : M \to T_s^r(M)$ be a smooth (r, s)-tensor field on $M, p \in M, q = X(\tau, p)$. Then $Z(p) \in T_s^r(T_pM), Z(X(\tau, p)) \in T_s^r(T_qM)$, and $A(\tau) = DX_s^r(\tau, p)^{-1}Z(X(\tau, p)) \in T_s^r(T_pM)$, so $A(\tau)$ is a smooth curve of (r, s)-tensors in the fixed tensor space $T_s^r(T_pM)$. The Lie derivative of Z in the direction of U (or along U) is denoted by $[Z, U]$ and is defined as

$$[Z, U](p) = \frac{\partial}{\partial\tau}A(\tau)\,|_{\tau=0} = \frac{\partial}{\partial\tau}DX_s^r(\tau, p)^{-1}Z(X(\tau, p))\,|_{\tau=0}. \tag{5}$$

Since $A(\tau) \in T_s^r(T_pM)$ for all τ its derivative is in $T_s^r(T_pM)$ so $[Z, U](p) \in T_s^r(T_pM)$ and $[Z, U]$ is a smooth (r, s)-tensor field also and $[\cdot, \cdot] : T_s^r(M) \times T(M) \to T_s^r(M)$ is bilinear. $[\cdot, \cdot]$ is called the Lie bracket.

If M is an open set in $\mathbf{R}^m$ and $Z : M \to \mathbf{R}$ is a smooth function $((0,0)$-tensor field) then in classical notation

$$[Z, U](x) = \nabla Z(x) \cdot U(x) \tag{6}$$

so $[Z, U]$ is the directional derivative of Z in the direction U. If Z is a smooth vector field (ordinary differential equation) as in (2) then

$$[Z, U](x) = \frac{\partial U}{\partial x}(x)Z(x) - \frac{\partial Z}{\partial x}(x)U(x) \tag{7}$$

where Z and U are column vectors. If Z is a one form thought of as a column vector then

$$[Z, U](x) = \frac{\partial U}{\partial x}(x)^T Z(x) + \frac{\partial Z}{\partial x}(x)^T U(x). \tag{8}$$

Suppose that the perturbation problem is given as an (r, s)-tensor field $Z = Z_*$ on M which has a formal expansion in a small parameter ϵ. Consider

$$Z(\epsilon, x) = Z_*(\epsilon, x) = \sum_{j=0}^{\infty} \left(\frac{\epsilon^j}{j!}\right) Z_j^0(x) \tag{9}$$

where each $Z_j^0 : M \to T_s^r M$ is an (r,s)-tensor field.

To simplify the perturbation problem given by Z_* in (9) we seek a near identity change of coordinates of the form

$$x = X(\epsilon, y) = y + \cdots \qquad (10)$$

where $X(\epsilon, y)$ is constructed as a formal solution of the non autonomous system of differential equations

$$\frac{dx}{d\epsilon} = W(x, \epsilon) = \sum_{j=0}^{\infty} \left(\frac{\epsilon^j}{j!}\right) W_{j+1}(x), \qquad (11)$$

satisfying the initial condition

$$x(0) = y \qquad (12)$$

where each $W_j : M \to TM$ is a smooth vector field.

The Lie transform of $Z(= Z_*)$ by W, denoted by $\mathcal{L}(W)Z$ or Z^*, is the tensor field Z_* expressed in the new coordinates and so is an (r,s)-tensor field depending on the parameter ϵ also. Specifically,

$$Z^*(\epsilon, y) = \mathcal{L}(W)Z(\epsilon, y) = DX_s^r(\epsilon, y)^{-1} Z_*(\epsilon, X(\epsilon, y)). \qquad (13)$$

In the new coordinates y the tensor $Z_*(x, \epsilon)$ becomes

$$Z^*(y, \epsilon) = \mathcal{L}(W)Z(\epsilon, y) = \sum_{j=0}^{\infty} \left(\frac{\epsilon^j}{j!}\right) Z_0^j(y). \qquad (14)$$

We say (10) or (11) transforms (9) into (14). The method of Lie transforms introduces a double indexed array of tensor fields $\{Z_j^i\}, i, j = 0, 1, \ldots$, which agree with the definitions given in (9) and (14) when either i or j is zero. The other terms are intermediary terms introduced to facilitate the computation. The main theorem on Lie transforms by Deprit(1969) in this general context is the following.

Deprit's Theorem. *Using the notation given above, the tensor fields* $\{Z_j^i\}, i = 1, 2, \ldots, j = 0, 1, \ldots,$ *satisfy the recursive identities*

$$Z_j^i = Z_{j+1}^{i-1} + \sum_{k=0}^{j} \binom{j}{k} [Z_{j-k}^{i-1}, W_{k+1}]. \qquad (15)$$

Remarks. The above formulas contain the standard binomial coefficient $\binom{j}{k} = \frac{j!}{k!(j-k)!}$. Note that since the transformation generated by W is a near identity transformation the first terms in Z_* and Z^* are the same, namely Z_0^0. Also note that the first term in the expansion for W starts with W_1. This convention imparts some nice properties to the formulas in (15). Each term in (15) has indices summing to $i + j$ and each term on the right hand side has upper index $i - 1$.

The interdependence of the $\{Z_j^i\}$ can easily be understood by considering the Lie triangle

$$
\begin{array}{ccccc}
Z_0^0 & & & & \\
\downarrow & & & & \\
Z_1^0 & \rightarrow & Z_0^1 & & \\
\downarrow & & \downarrow & & \\
Z_2^0 & \rightarrow & Z_1^1 & \rightarrow & Z_0^2 \\
\downarrow & & \downarrow & & \downarrow
\end{array}
\tag{16}
$$

The coefficients of the expansion of the old tensor field Z_* are in the left column and those of the new tensor field Z^* are on the diagonal. The formula (15) says that to calculate any element in the Lie triangle you need the entries in the column one step to the left and up.

Proof. Let $Y(\epsilon, x)$ be the inverse of $X(\epsilon, y)$ so
$Y(\epsilon, X(\epsilon, y)) \equiv y$, $X(\epsilon, Y(\epsilon, x)) \equiv x$,
$DX(\epsilon, y)^{-1} = DY(\epsilon, X(\epsilon, y))$, and $DX_s^r(\epsilon, y)^{-1} = DY_s^r(\epsilon, X(\epsilon, y))$.
Thus (13) becomes
$\mathcal{L}(W)Z(\epsilon, y) = DY_s^r(\epsilon, X(\epsilon, y))Z_*(\epsilon, X(\epsilon, y))$.

Define the differential operator $\mathcal{D} = \mathcal{D}_W$ acting on (r, s)-tensor fields depending on a parameter ϵ by

$$
\mathcal{D}K(\epsilon, x) = \frac{\partial K}{\partial \epsilon}(\epsilon, x) + [K, W](\epsilon, x).
\tag{17}
$$

In computing the Lie bracket in (17) the ϵ is simply a parameter and so held fixed during any differentiation. With this notation we have

$$
\frac{d}{d\epsilon}\left\{ DY_s^r(\epsilon, x)K(\epsilon, x) \mid_{x=X(\epsilon, y)} \right\} = DY_s^r(\epsilon, x)\mathcal{D}K(\epsilon, x) \mid_{x=X(\epsilon, y)}.
\tag{18}
$$

Define new functions by $Z^0 = Z, Z^i = \mathcal{D}Z^{i-1}, i \geq 1$. Let these functions have series expansions

$$
Z^i(\epsilon, x) = \sum_{k=0}^{\infty} \left(\frac{\epsilon^k}{k!} \right) Z_k^i(x),
\tag{19}
$$

so

$$Z^i(\epsilon, x) = D \sum_{k=0}^{\infty} \left(\frac{\epsilon^k}{k!}\right) Z_k^{i-1}(x)$$

$$= \sum_{k=0}^{\infty} \left(\frac{\epsilon^{k-1}}{(k-1)!}\right) Z_k^{i-1}(x) + \sum_{k=0}^{\infty} [\left(\frac{\epsilon^k}{k!}\right) Z_k^{i-1}(x), \sum_{s=0}^{\infty} \left(\frac{\epsilon^s}{s!}\right) W_{s+1}(x)]$$

$$= \sum_{j=0}^{\infty} \left(\frac{\epsilon^j}{j!}\right) \left(Z_{j+1}^{i-1} + \sum_{k=0}^{j} \binom{j}{k} [Z_{j-k}^{i-1}, W_{k+1}]\right).$$

(20)

So the functions Z_j^i are related by (15). It remains to show that $Z = G$ has the expansion (14). By Taylor's theorem and (18)

$$G(\epsilon, y) = \sum_{n=0}^{\infty} \left(\frac{\epsilon^n}{n!}\right) \frac{d^n}{d\epsilon^n} G(\epsilon, y) \mid_{\epsilon=0}$$

$$= \sum_{n=0}^{\infty} \left(\frac{\epsilon^n}{n!}\right) \frac{d^n}{d\epsilon^n} \left(DY_s^r(\epsilon, x) Z(\epsilon, x) \mid_{x=X(\epsilon, y)}\right)_{\epsilon=0}$$

(21)

$$= \sum_{n=0}^{\infty} \left(\frac{\epsilon^n}{n!}\right) \left(DY_s^r(\epsilon, x) \mathcal{D}^n Z(\epsilon, x) \mid_{x=X(\epsilon, y)}\right)_{\epsilon=0}$$

$$= \sum_{n=0}^{\infty} \left(\frac{\epsilon^n}{n!}\right) Z_0^n(x).$$

In the cases of interest the tensor field is given and the change of variables is sought to simplify it. When the field is sufficiently simple it is said to be in 'normal form'. The main Lie transform algorithm starts with a given field which depends on a small parameter, ϵ, and constructs a change of variables so that the field in the new variables is simple. The algorithm is built around the following observation.

Consider the series (9) as given so all the Z_i^0 are known. Assume that all the entries in the Lie triangle are known down to the N row, so the Z_j^i are known for $i + j \leq N$ and assume the W_i are known for $i \leq N$. Let $\tilde{Z}_j^i$ be computed from the same differential equation, so $\tilde{Z}_i^0 = Z_i^0$ for all i, and with $\mathcal{W}_1, \ldots, \mathcal{W}_N$ where $\mathcal{W}_i = W_i$ for $i = 1, 2, \ldots, N-1$ but $\mathcal{W}_N = 0$; then

$$\tilde{Z}_j^i = Z_j^i \quad \text{for} \quad i + j < N$$

(22)

$$\tilde{Z}_j^i = Z_j^i + [Z_0^0, W_N] \quad \text{for} \quad i + j = N.$$

This is easily seen from the recursive formulas in (15). Recall the remark that the sum of all the indices must add to the row number, so W_N does not affect the terms in the first $N-1$ rows. The second equation in (22) follows from a simple induction across the Nth row. The algorithm can be used to prove a general theorem which includes almost all applications, see Meyer and Schmidt(1977).

General Normalization Theorem. *Let $\{\mathcal{P}_i\}_{i=0}^{\infty}, \{\mathcal{Q}_i\}_{i=1}^{\infty}$ and $\{\mathcal{R}_i\}_{i=1}^{\infty}$ be sequences of linear spaces of smooth fields defined on a manifold M where $\{\mathcal{P}_i\}_{i=0}^{\infty}$ and $\{\mathcal{Q}_i\}_{i=1}^{\infty}$ are (r,s)-tensor fields and $\{\mathcal{R}_i\}_{i=1}^{\infty}$ are vector fields. Assume:*

i) $\mathcal{Q}_i \subset \mathcal{P}_i$, $i = 1, 2, \ldots;$

ii) $Z_i^0 \in \mathcal{P}_i$, $i = 0, 1, 2, \ldots;$

iii) $[\mathcal{P}_i, \mathcal{R}_j] \subset \mathcal{P}_{i+j}$, $i, j = 0, 1, 2, \ldots;$

iv) for any $A \in \mathcal{P}_i, i = 1, 2, \ldots$, there exist $B \in \mathcal{Q}_i$ and $C \in \mathcal{R}_i$ such that

$$B = A + [Z_0^0, C]. \tag{3}$$

Then there exists a W with a formal expansion of the form (11) with $W_i \in \mathcal{R}_i, i = 1, 2, \ldots$, which transforms the tensor field Z_* with the formal series expansion given in (9) to the field Z^* with the formal series expansion given by (14) with $Z_0^i \in \mathcal{Q}_i, i = 1, 2, \ldots.$

Proof. Use induction on the rows of the Lie triangle. Induction Hypothesis I_n: Let $Z_j^i \in \mathcal{P}_{i+j}$ for $0 \le i + j \le n$ and $W_i \in \mathcal{R}_i$, $Z_0^i \in \mathcal{Q}_i$ for $1 \le i \le n$.

I_0 is true by assumption and so assume I_{n-1}. By (15)

$$Z_{n-1}^1 = Z_n^0 + \sum_{k=0}^{n-2} \binom{n-1}{k} [W_{k+1}, Z_{n-1-k}^0] + [W_n, Z_0^0]. \tag{24}$$

The last term is singled out because it is the only term that contains an element, W_n, which is not covered either by the induction hypothesis or the hypothesis of the theorem. All the other terms are in $\mathcal{P}_n$ by I_{n-1} and iii). Thus

$$Z_{n-1}^1 = K^1 + [W_n, Z_0^0] \tag{25}$$

where $K^1 \in \mathcal{P}_n$ is known. A simple induction on the columns of the Lie triangle using (15) shows that

$$Z_{n-s}^s = K^s + [W_n, Z_0^0] \tag{26}$$

where $K^s \in \mathcal{P}_n$ for $s = 1, 2, \ldots, n$ and so

$$Z_n^0 = K^n + [W_n, Z_0^0]. \tag{27}$$

By iv) solve (27) for $W_n \in \mathcal{R}_n$ and $Z_0^n \in \mathcal{Q}_i$. Thus I_n is true.

The theorem given above is formal in the sense that the convergence of the various series is not discussed. In interesting cases the series diverge, but useful information can be obtained in the first few terms of the normal form. One can stop the process at any order, N, to obtain a W which is a polynomial in ϵ and so converges. From the proof given above it is clear that the terms in series for Z^* up to order N are unaffected by the termination.

6.   Function applications

In this section I will show some applications of the method of Lie transforms when the problem involves simply functions as opposed to vector fields.

The implicit function theorem. One of the fundamental theorems of analysis is the implicit function theorem. I will show how to compute the implicitly defined function using Lie transforms.

Consider a real valued analytic function (or formal power series) $f(x, y)$ defined in neighborhood of the origin in $\mathbf{R}^2$ such that $f(0, 0) = 0$ and $\frac{\partial f}{\partial y}(0, 0) = \alpha \neq 0$. Then the implicit function theorem asserts that there is an analytic function (or formal power series) $\psi(x)$ such that $\psi(0) = 0$ and $f(x, \psi(x)) \equiv 0$. Introduce a small parameter ϵ by scaling $x \to \epsilon^2 x, y \to \epsilon y$ and $f \to \epsilon^{-1} f$, that is define F_* by

$$F_*(x, y, \epsilon) = \epsilon^{-1} f(\epsilon^2 x, \epsilon y) = \sum_{i=0}^{\infty} \left(\frac{\epsilon^i}{i!} \right) F_i^0(x, y) \tag{1}$$

and $F_0^0(x, y) = \alpha y$. Let y be the variable and treat x simply as a parameter in the problem. The functions $F_i^0(x, y)$ are polynomials in x and y and so let $\mathcal{P}_i$ be the vector space of polynomials in x and y.

By the Normalization Theorem we must be able to solve the Lie equation (7.20) where A is any polynomial. In this case the Lie bracket is $[C, F_0^0] = \frac{\partial C}{\partial y} \alpha$. Clearly we can solve $[C, F_0^0] + A = B$ by taking $B = 0$ and by taking C as an indefinite integral of $-\alpha^{-1} A$. Thus if we define $\mathcal{Q}_i = \{0\}$ and $\mathcal{R}_i = \mathcal{P}_i$, then for any $A \in \mathcal{P}_i$, we can solve (18) for $B \in \mathcal{Q}_i = \{0\}$ and $C \in \mathcal{R}_i = \mathcal{P}_i$. Thus one can compute a transformation such that $F^*(x, y, \epsilon) = \alpha y$. But $F^*(x, y, \epsilon) = F_*(x, \phi(x, y, \epsilon), \epsilon) = \epsilon^{-1} f(\epsilon^2 x, \epsilon \phi(x, y, \epsilon))$. So $\phi(x, 0, 1) = \psi(x)$ satisfies $f(x, \psi(x)) \equiv 0$. This shows that the implicit function can be computed by Lie transforms. In general the method of Lie transforms only produces a formal series, but in this case the implicit function theorem ensures that formal series converges when the series for f does. The general vector form of the implicit function theorem can be handled in a similar manner.

The splitting lemma. The splitting lemma is an important tool in the analysis of critical points of a function and catastrophe theory (see Poston and Stewart(1978)). Let $V(x)$ be a real valued analytic function defined in a neighborhood of the origin in $\mathbf{R}^m$ and $x \in \mathbf{R}^m$. Assume that the origin is a critical point for V and for simplicity assume that $V(0) = 0$. Assume that the rank of the Hessian, $\partial^2 V(0)/\partial x^2$, is s, $0 \leq s \leq m$. Then the splitting lemma says that there is a change of coordinates $x = \phi(y)$ such that in the new coordinates

$$V(y) = (\pm y_1^2 \pm \cdots \pm y_s^2)/2 + v(y_{s+1}, ..., y_m). \tag{2}$$

Scale by $x \to \epsilon x$, and $V \to \epsilon^{-2}V$, or define

$$U_*(x, \epsilon) = \epsilon^{-2}V(\epsilon x) = \sum_{i=0}^{\infty} \left(\frac{\epsilon^i}{i!}\right) U_i^0(x). \tag{3}$$

Here the $U_i^0(x, \mu)$ are polynomials in x of degree $i + 2$, so let $\mathcal{P}_i$ be the vector space of such polynomials. $U_0^0(x)$ is a quadratic form in x and so by making a linear change of variable if necessary we may assume that

$$U_0^0(x) = (\pm x_1^2 \pm x_2^2 \pm \cdots \pm x_s^2)/2. \tag{4}$$

To solve the Lie equation let

$$C = c x_1^{k_1} \cdots x_m^{k_m} \tag{5}$$

be monomial of degree $i + 2$ and where $c = (c_1, \ldots, c_m)^T$ is an n-vector. Then

$$[C, U_0^0] = \pm c_1 x_1^{k_1+1} x_2^{k_2} \cdots x_m^{k_m} \pm \cdots \pm c_s x_1^{k_1} x_2^{k_2} \cdots x_s^{k_s+1} \cdots x_m^{k_m} \tag{6}$$

so the kernel of $[\cdot, U_0^0]$ consists of all homogeneous polynomials of degree $i+2$ in $x_s, \ldots, x_m$ and the range of $[\cdot, U_0^0]$ consists of the span of all monomials which contain one of $x_1, \ldots, x_s$ to a positive power or equivalently those polynomials which are zero when $x_1 = \cdots = x_s = 0$. Thus we can solve the Lie equation by taking $\mathcal{P}_i$ as the space of all scalar homogeneous polynomials of degree $i+2$, $\mathcal{Q}_i$ the subspace of $\mathcal{P}_i$ consisting of all scalar homogeneous polynomials of degree $i + 2$ in $x_s, \ldots, x_m$ alone, and $\mathcal{R}_i$ the space of all n-vectors of homogeneous polynomials of degree $i + 1$ in $x_1, \ldots, x_m$.

Thus the method of Lie transforms will construct a change of coordinates so that in the new coordinates

$$U^*(y, \epsilon) = \sum_{i=0}^{\infty} \left(\frac{\epsilon^i}{i!}\right) U_0^i(y) \tag{7}$$

where for $i \geq 1$ the $U_0^i(y)$ depend only on $y_s, \ldots, y_n$. Setting $\epsilon = 1$ gives the form given by the splitting lemma in (2).

In Meyer and Schmidt(1987) the problem for finding bifurcations of relative equilibria in the N-body problem was reduced to finding the bifurcation of critical points of the potential constrained to a constant moment of inertia manifold. The constraint equation was solved by the method of Lie transforms to compute the implicitly defined function. Then by applying the splitting lemma algorithm we obtained the bifurcation equations in a form that could be analyzed by hand.

7. Autonomous differential equations

In this section I will show how the Normalization Theorem can be used to study autonomous differential equations. There are many more applications than the ones given here.

The classical normal form. This is the case discussed in Sections 2 and 3. Consider the equation

$$\dot{x} = Ax + f(x) \tag{1}$$

where $x \in \mathbf{R}^m$, A is an $m \times m$ constant matrix, f is an analytic function defined in a neighborhood of the origin in $\mathbf{R}^m$ whose series expansion starts with second degree terms. Scale the equations by $x \to \epsilon x$ and divide the equation by ϵ so that (1) becomes

$$\dot{x} = \sum_{i=0}^{\infty} \left(\frac{\epsilon^i}{i!} \right) F_i^0(x), \tag{2}$$

where $F_0^0(x) = Ax$ and F_i^0 is an m-vector of homogeneous polynomials of degree $i + 1$ so let $\mathcal{P}_i$ be the space of all such polynomials.

Assume that A is diagonal so $A = \operatorname{diag}(\lambda_1, \dots, \lambda_m)$. In order to solve (18) let

$$A = ax^k, \qquad B = bx^k, \qquad C = cx^k,$$

$$k = (k_1, \dots, k_m), \quad x = (x_1, \dots, x_m), \quad x^k = x_1^{k_1} \cdots x_m^{k_m}, \tag{3}$$

and substitute into the Lie equation to get

$$bx^k = ax^k + \left(A - (\Sigma k_s \lambda_s) I \right) cx^k. \tag{4}$$

The coefficient matrix, $A - (\Sigma k_s \lambda_s) I$, of cx^k is diagonal with entries $\lambda_j - \Sigma k_s \lambda_s$. So to solve (4) take

$$c_j = \frac{-a_j}{\lambda_j - \Sigma k_s \lambda_s}, \ b_j = 0 \text{ when } \lambda_j - \Sigma k_s \lambda_s \neq 0,$$

$$c_j = 0, b_j = a_j \text{ when } \lambda_j - \Sigma k_s \lambda_s = 0. \tag{5}$$

Let $e_j = (0, \dots, 0, 1, 0, \dots, 0)^T$ be the standard basis for $\mathbf{R}^m$. From the above we define

$$\mathcal{Q}_i = \operatorname{span}\{e_j x^k : \lambda_j - \Sigma k_s \lambda_s = 0, \Sigma k_s = i + 1\},$$

$$\mathcal{R}_i = \operatorname{span}\{e_j x^k : \lambda_j - \Sigma k_s \lambda_s \neq 0, \Sigma k_s = i + 1\}, \tag{6}$$

so the condition in ii) of the main theorem is satisfied. So (2) can be formally transformed to

$$\dot{y} = \sum_{i=0}^{\infty} \left(\frac{\epsilon^i}{i!} \right) F_0^i(y), \tag{7}$$

where $F_0^i \in \mathcal{Q}_i$ for all $i \geq 1$. Setting $\epsilon = 1$ bring the equations to the form

$$\dot{y} = Ay + g(y) \tag{8}$$

where the terms in g lie in some $\mathcal{Q}_i$. It is easy to check that a term $h(y)$ is in some $\mathcal{Q}_i$ if and only if $h(e^{At}y) = e^{At}h(y)$ for all y and t. Thus g in (8) satisfies

$$g(e^{At}y) = e^{At}g(y). \tag{9}$$

This formulation for the normal form does not require that A be in diagonal form (A must be diagonalizable!). This proves the Classical Formal Theorem and the Classical Finite Normalization Theorem of Section 3.

The general equilibria. Recently there has been a lot of progress on normal forms in the case when A is not diagonalizable or simple, and the research goes on. First Kummer (1976,1978) and then Cushman, Deprit and Mosak (1983), used group representation theory to study the normal forms. Representation theory is very helpful in understanding the general case, but there are simpler ways to understand the basic ideas and examples. In Meyer(1984) a theorem like the Classical Normalization Theorem of Section 3 with A was replaced by A^T was given; so, the terms in the normal form are invariant under the flow $\exp(A^T t)$. A far better proof can be found in Elphick et al. (1987), which is what I will outline here.

The proof in the classical case rested on the fact that for a simple matrix, A, $\mathbf{R}^m$ is the direct sum of the range and kernel of A, and this held true for the Lie operator $L = [\cdot, Ax]$ defined on homogeneous polynomials as well. The method of Elphick et al. is based on the following simple lemma in linear algebra known as the Fredholm alternative and an inner product defined on homogeneous polynomials given after the lemma.

Fredholm Alternative. *Let* $\mathbf{V}$ *be a finite dimensional inner product space with inner product* $(\cdot, \cdot)$. *Let* $A : \mathbf{V} \to \mathbf{V}$ *be a linear transformation, and* A^* *its adjoint (so* $(Ax, y) = (x, A^*y)$ *for all* $x, y \in \mathbf{V}$). *Then* $\mathbf{V} = R \oplus K^*$ *where* R *is the range of* A *and* K^* *is the kernel of* A^*.

Proof. Exercise.

Let $\mathcal{P} = \mathcal{P}_i$ be the linear space of all m-vectors of homogeneous polynomials of degree i in m variables $x \in \mathbf{R}^m$. So if $P \in \mathcal{P}$, then

$$P(x) = \sum_{|k|=i} p_k x^k = \sum_{|k|=i} p_{k_1 k_2 \ldots k_m} x_1^{k_1} x_2^{k_2} \cdots x_m^{k_m}, \qquad (10)$$

where the $p_k \in \mathbf{R}^m$. Define $P(\partial)$ to be the differential operator

$$P(\partial) = \sum_{|k|=i} p_k \frac{\partial^k}{\partial x^k}, \qquad (11)$$

where we have introduced the notation

$$\frac{\partial^k}{\partial x^k} = \frac{\partial^{k_1}}{\partial x_1^{k_1}} \frac{\partial^{k_2}}{\partial x_2^{k_2}} \cdots \frac{\partial^{k_m}}{\partial x_m^{k_m}}. \qquad (12)$$

Let $Q \in \mathcal{P}, Q(x) = \sum q_h x^h$ be another homogeneous polynomial, and define an inner product $< \cdot, \cdot >$ on $\mathcal{P}$ by

$$< P, Q > = P(\partial) \cdot Q(x). \qquad (13)$$

To see that this is indeed an inner product, note that $\partial^k x^h / \partial x^k = 0$ if $k \neq h$ and $\partial^k x^k / \partial x^k = k! = k_1! k_2! \cdots k_{2n}!$ if $k = h$; so,

$$< P, Q > = \sum_{|k|=i} k! p_k \cdot q_k. \qquad (14)$$

Let A be a general $m \times m$ matrix. The Lie operator is $L_A : \mathcal{P} \to \mathcal{P}$, where

$$L_A P = [P, Ax] = AP - \frac{\partial P}{\partial x} Ax = \frac{d}{dt} e^{At} P(e^{-At} x) \mid_{t=0}. \qquad (15)$$

Lemma. *Let $A : \mathbf{R}^m \to \mathbf{R}^m$ be as above and A^T its transpose (so A^T is the adjoint of A with respect to the standard inner product in $\mathbf{R}^m$). Then for all $P, Q \in \mathcal{P}$,*

$$< P, L_A Q > = < L_{A^T} P, Q > . \qquad (16)$$

That is, the adjoint of L_A with respect to $< \cdot, \cdot >$ is L_{A^T}.

Proof. The proof is not hard – see Elphick et al.(1987) or Meyer and Hall(1992).

General Equilibrium Theorem. *Let A be an $m \times m$ matrix and A^T its transpose. Then there exists a formal change of variables, $x = X(y) = y + \cdots$, which transforms the equation (1) to (8) where*

$$g(e^{A^T t} y) = e^{A^T t} g(y), \qquad (17)$$

or equivalently

$$[g, A^T x] = 0. \tag{18}$$

Proof. We must solve the Lie equation $L_A C + D = B$, where $D \in P_i = P$ is given, and $C \in Q_i = P$, and $D \in Q_i = kernel(L_A T)$. By the lemma above, we can write $D = B - G$, where $B \in kernel(L_A T)$; so, $[B, A^T x] = 0$, and $G \in range(L_A)$; so, $G = L_A C, C \in P$. With these choices the Lie equation is solved. Verification of the rest of the hypothesis in the general perturbation theorem is straightforward.

An example of normal forms in the non-simple case. Consider the equation (1) with $m = 1$ and

$$A = \begin{pmatrix} 0 & 1 \\ 0 & 0 \end{pmatrix}, \quad A^T = \begin{pmatrix} 0 & 0 \\ 1 & 0 \end{pmatrix}. \tag{19}$$

Since $\exp(A^T t) = \begin{pmatrix} 1 & 0 \\ t & 1 \end{pmatrix}$ equation (17) implies that the system is in normal form if

$$\dot{y}_1 = \alpha(y_1) y_1 + y_2,$$

$$\dot{y}_2 = \beta(y_1) + \alpha(y_1) y_2. \tag{20}$$

In the case when

$$A = \begin{pmatrix} i & 1 & 0 & 0 \\ 0 & i & 0 & 0 \\ 0 & 0 & -i & 1 \\ 0 & 0 & 0 & -i \end{pmatrix} \tag{21}$$

the normal form is

$$\dot{y}_1 = (i + h(a,b)) y_1 + y_2,$$

$$\dot{y}_2 = k(a,b) y_1 + (i + h(a,b)) y_2,$$

$$\dot{y}_3 = (-i + \bar{h}(\bar{a},\bar{b})) y_3 + y_4,$$

$$\dot{y}_4 = \bar{k}(\bar{a},\bar{b}) y_3 + (-i + \bar{h}(\bar{a},\bar{b})) y_4, \tag{22}$$

where $a = y_1 y_3$ and $b = y_1 y_4 - y_3 y_2$. The normal form given in Meyer(1984) for this last case is not quite correct.

8. Non-autonomous differential equations

In many applications the differential equations involve time explicitly so one must consider equations of the form $\dot{x} = f(t,x)$. In this case one would allow the transformation generated by W to depend on t also. But this case can be reduced to the previous case by replacing the original system with the equivalent autonomous system $\dot{x} = f(\tau, x), \dot{\tau} = 1$ where τ is a new variable.

Consider the system

$$\dot{x} = Z_*(t, x, \epsilon) = \sum_{j=0}^{n} \left(\frac{\epsilon^j}{j!} \right) Z_j^0(t, x), \tag{1}$$

and the near identify transformation

$$x = x(t, \xi, \epsilon) = \xi + \cdots \tag{2}$$

generated as a solution of the equation

$$\frac{dx}{d\epsilon} = W(t, x, \epsilon) = \sum_{j=0}^{n} \left(\frac{\epsilon^j}{j!} \right) W_{j+1}(t, x), x(0) = \xi \tag{3}$$

which transforms (43) to

$$\dot{\xi} = Z^*(t, \xi, \epsilon) = \sum_{j=0}^{n} \left(\frac{\epsilon^j}{j!} \right) Z_0^j(t, \xi). \tag{4}$$

The translation of the main theorem to the non-autonomous case goes as follows.

Time Dependent Normalization Theorem. *Let $\mathcal{P}_j$ and $\mathcal{R}_j$ be linear spaces of smooth time dependent vectors fields defined for $(t, x) \in D \subset \mathbf{R}^{m+1}$ and let $\mathcal{Q}_j$ be a subspace of $\mathcal{P}_j$. If i)$Z_j^0 \in \mathcal{P}_j$ for $j = 0, 1, 2, ...,$ ii)$[\mathcal{P}_i, \mathcal{Q}_j] \subset \mathcal{P}_{i+j}, i, j = 0, 1, 2, ...,$ iii) for any $i = 1,2,3,...$ and any $A \in \mathcal{P}_i$ there exist $B \in \mathcal{Q}_i$ and $C \in \mathcal{R}_i$ such that*

$$B = A + [C, Z_0^0] - \dot{C}, \tag{5}$$

then one can construct W as in (3) with $W_i \in \mathcal{R}_i$ which generates a transformation (2) which takes (1) to (4) with $Z_0^i \in \mathcal{Q}_i$.

The method of averaging. The method of averaging is a special case of the normal form theorem given above. The method of averaging deals with a periodic system of the form (43) where $Z_0^0 = 0$, i.e. $\dot{x} = \epsilon Z_1^0(t, x) + \cdots$. One seeks a periodic change of variables, so the function W must be periodic in t also. Equation (5) reduces to $B = A + \dot{C}$. Given a periodic A in order to have a periodic C it is necessary and sufficient that we take B as the average over a period of A, so B is independent of t, and C as any indefinite integral of $B - A$. This shows that the normalized equations (4) are autonomous, i.e. Z_0^i is independent of t. The name comes from the fact that Z_0^1 is the time average of Z_1^0.

The Floquet exponents and the Liapunov transformation. A classical problem is to compute the characteristic exponents of Mathieu's equation $\ddot{y} + (a + b\cos 2\pi t)x = 0$ or other similar linear periodic systems. Assume that $Z_0^0(t, x) = Ax$ where A is a diagonal matrix $A = diag(\lambda_1, \ldots, \lambda_m)$ and $Z_i^0(t, x) = A_i(t)x$ where $A_i(t)$ is an $n \times nT$-periodic matrix, so let $\mathcal{P}_i$ be the space of all linear T-periodic systems. Seek a linear T-periodic change of variables, so seek $W_i(t, x) = C_i(t)x$ where $C_i(t)$ is to be T-periodic also. Equation (5) becomes

$$B(t) = A(t) + C(t)A - AC(t) - \dot{C}(t) \tag{6}$$

where A, B and C are matrices. The equation for the ij^{th} component is

$$b_{ij} = a_{ij} + (\lambda_i - \lambda_j)c_{ij} - \dot{C}_{ij}. \tag{7}$$

This is a linear first order differential equation in c_{ij}. If $(\lambda_i - \lambda_j)T \neq n2\pi i$ then take b_{ij} to be the average of a_{ij} and c_{ij}, the unique T-periodic solution of (7). Thus the space $\mathcal{Q}_i$ is all linear systems with constant diagonal coefficient matrices. Thus we can compute a linear periodic change of coordinates which reduces the linear periodic system (1) to the linear diagonal constant system (4), this transformation is known as the Liapunov transformation. The entries on the diagonal are the Floquet exponents. The equation (6) has been studied in the more general case when A is not necessarily diagonal. The presentation given here is merely a simple example.

A very similar problem is to calculate the series expansion of a solution of a linear differential equation at a regular singular point.

9. Classical Hamiltonian systems

For a much more complete discussion of normal forms for Hamiltonian systems read Meyer and Hall(1992). This section is just a taste.

For Hamiltonian systems the Lie bracket is replaced by the Poisson bracket. Let F, G and H be smooth real valued functions defined in an open set in $\mathbf{R}^{2n}$, the Poisson bracket of F and G is the smooth function $\{F, G\}$ defined by

$$\{F, G\} = \frac{\partial F^T}{\partial x} J \frac{\partial G}{\partial x} \tag{1}$$

where J is the usual $2n \times 2n$ skew symmetric matrix of Hamiltonian mechanics. A Hamiltonian differential equation (generated by the Hamiltonian H) is

$$\dot{x} = J\frac{\partial H}{\partial x}. \tag{2}$$

The Poisson bracket and the Lie bracket are related by

$$J\frac{\partial}{\partial x}\{F,G\} = \left[J\frac{\partial F}{\partial x}, J\frac{\partial G}{\partial x}\right] \qquad (3)$$

so the Hamiltonian vector field generated by $\{F,G\}$ is the Lie bracket of the Hamiltonian vector fields generated by G and F, see Abraham and Marsden(1978).

Consider a Hamiltonian perturbation problem given by the Hamiltonian

$$H_*(x,\epsilon) = \sum_{j=0}^{n}\left(\frac{\epsilon^j}{j!}\right) H_j^0(x). \qquad (4)$$

A near identity symplectic change of coordinates $x = \phi(\xi,\epsilon) = \xi + \cdots$ can be generated as the solution of the Hamiltonian differential equations

$$\frac{dx}{d\epsilon} = J\frac{\partial W}{\partial x}(x,\epsilon), x(0) = \xi, W(x,\epsilon) = \sum_{j=0}^{n}\left(\frac{\epsilon^j}{j!}\right) W_{j+1}(x). \qquad (5)$$

It transforms (4) to

$$H^*(x,\epsilon) = \sum_{j=0}^{n}\left(\frac{\epsilon^j}{j!}\right) H_0^j(x). \qquad (6)$$

Using these facts given above one can translate the Normalization Theorem to:

The Hamiltonian Normalization Theorem. *Let $\mathcal{P}_j$, $\mathcal{Q}_j$, and $\mathcal{R}_j$ be vector spaces of smooth Hamiltonians on D with $\mathcal{Q}_j \subset \mathcal{P}_j$. Assume that i) $Z_j^0 \in \mathcal{P}_j$ for $j = 1,2,3...,$ ii) $\{\mathcal{P}_i, \mathcal{Q}_j\} \subset \mathcal{P}_{i+j}$ for $i,j = 1,2,3,...,$ iii) for any i and any $A \in \mathcal{P}_j$ there exist $B \in \mathcal{Q}_j$ and $C \in \mathcal{R}_j$ such that*

$$B = A + \{C, H_0^0\}. \qquad (7)$$

Then one can compute a formal expansion for W in (5) with $W_j \in \mathcal{R}_j$ for all j which transforms (4) to (6) where $H_0^j \in \mathcal{Q}_j$ for all j.

The classical Birkhoff normal form for a Hamiltonian system near an equilibrium point is as follows. Assume that the Hamiltonian (4) came from scaling a system about an equilibrium point at the origin. That is, $H_0^0(x)$ is a quadratic form and H_j^0 is a homogeneous polynomial of degree $j + 2$. Assume that the linear Hamiltonian system

$$\dot{x} = J\frac{\partial H_0^0}{\partial x} = Ax \qquad (8)$$

is such that A is diagonalizable. Then one can compute a symplectic change of variables generated by (5) which transforms (4) to (6) with

$$H^*(e^{At}x, \epsilon) = H^*(x, \epsilon). \tag{9}$$

For a Lie transform proof see Meyer(1974).

Kummer(1976,1978) has shown that Lie algebra theory is useful in studying normal forms in some special cases in celestial mechanics. Taking this lead Cushman, Deprit and Mosak(1983) have used results from representation theory to give a complete description of the normal forms for Hamiltonian systems without the diagonalizable assumption.

10. The computational Poincare Lemma and Darboux Theorem

To my knowledge the method of Lie transforms has not been used on tensor fields more complicated than vector fields. Here I will give an example to illustrate the generality of the method and in the next section show how to apply it. In order to avoid the notational overload found in modern treatises like Kobayashi and Nomizu(1963) or Abraham and Marsden(1978), I shall use classical tensor notation. Thus repeated indices are summed over. Since the problem is a computational one we must use coordinates in the end anyway. Flanders(1963) is a highly recommended introduction to differential forms. The fundamental geometry of Hamiltonian mechanics is embodied in a *symplectic structure*, Ω, i.e., a closed non-degenerate 2-form. In a neighborhood of the origin in $\mathbf{R}^{2n}$

$$\Omega = \Omega_{ij}(x)dx^i \wedge dx^j \tag{1}$$

where we have used the summation convention, $\Omega_{ij} = -\Omega_{ji}$, and the $\Omega_{ij}(x)$ are real analytic in x. $\{\Omega_{ij}\}$ is a 2-covariant tensor, so if you change coordinates by $x = x(y)$ then the tensor in the y coordinates is

$$\Omega(y) = \Omega_{ij}(x(y))\frac{\partial x^i}{\partial y_m}\frac{\partial x^j}{\partial y_n} dy^m \wedge dy^n. \tag{2}$$

Sometimes we will think of $\Omega(x)$ as the skew symmetric matrix $(\Omega_{ij}(x))$, the coefficient matrix of the form (1). Ω is non-degenerate means that the matrix $\Omega(x)$ is nonsingular for all x. (2) means that the matrix Ω transforms by

$$\Omega \rightarrow \frac{\partial x}{\partial y}^T \Omega \frac{\partial x}{\partial y}. \tag{3}$$

Ω is closed means that

$$d\Omega = \frac{\partial \Omega_{ij}}{\partial x_k} dx^i \wedge dx^j \wedge dx^k = 0. \tag{4}$$

Since we are working locally, a closed form is exact by Poincaré's Lemma so there is a one form $\alpha(x) = \alpha_i(x)dx^i$ such that $\Omega = d\alpha$. The computational form of Poincaré's Lemma is given below.

This matrix $\Omega(0)$ is nonsingular and skew symmetric so there is a nonsingular matrix P such that

$$P^T \Omega(0) P = J = \begin{pmatrix} 0 & I \\ -I & 0 \end{pmatrix}, \tag{5}$$

so by a linear change of coordinates the coefficient matrix of $\Omega(0)$ is J. Darboux's theorem says there is a nonlinear change of coordinates defined in a neighborhood of the origin in $\mathbf{R}^{2n}$ so that in the new coordinates the coefficient matrix of Ω is identically J in the whole neighborhood. Our computational procedure follows the proof given by Weinstein(1971).

Assume that the linear change of variables has been made so that $\Omega(0) = J$ and scale by $x \to \epsilon x, \Omega \to \epsilon^{-1}\Omega$ so that

$$\Omega = \sum_{s=0}^{\infty} \left(\frac{\epsilon^s}{s!}\right) \omega_s^0, \tag{6}$$

where ω_s^0 is a closed 2-form with coefficients that are homogeneous polynomials in x of degree s. Let $\mathcal{P}_s$ be the vector space of such forms and $\mathcal{Q}_s = \{0\}$. Let $A \in \mathcal{P}_s, B = 0 \in \mathcal{Q}_s$, and $C \in \mathcal{R}_s$, where $\mathcal{R}_s$ is the vector space of vector fields which are homogeneous polynomials of degree $s + 1$. In coordinates the Lie equation for this problem is

$$0 = A_{sm} + J_{im}\frac{\partial C^i}{\partial x_s} + J_{sj}\frac{\partial C^j}{\partial x_m}. \tag{7}$$

(In general there would be a term $+\partial J^{sm}/\partial x_i C^i$ in (7) but this term is zero since J is constant.)

Since A is a closed two form there is a one form α such that $A = d\alpha$ – see the Computational Poincaré Lemma given below. So (7) becomes

$$0 = \frac{\partial \alpha^s}{\partial x_m} - \frac{\partial \alpha^m}{\partial x_s} + J_{im}\frac{\partial C^i}{\partial x_s} + J_{sj}\frac{\partial C^j}{\partial x_m}. \tag{8}$$

This equation has a solution $C^i = \alpha_{i+n}$ for $1 \leq i \leq n$, $C^i = -\alpha_{i-n}$ for $n \leq i \leq 2n$, or $C = J\alpha$. Thus there is a solution of (18) and so the coordinate change given by Darboux's Theorem can be computed by Lie transforms.

Computational Poincare's Lemma. *Let $A = \sum A_{ij}(x)dx^i \wedge dx^j$ be a closed 2-form with each $A_{ij}(x)$ a homogeneous polynomial of degree p, then $A = dB$ where*

$$B = (2/p+2)\sum A_{ij}(x)x^i \wedge dx^j.$$

Proof. The form A is closed means

$$\frac{\partial A_{ij}}{\partial x^k} + \frac{\partial A_{ki}}{\partial x^j} + \frac{\partial A_{jk}}{\partial x^i} = 0. \tag{9}$$

Since each $A_{ij}(x)$ is a homogeneous polynomial of degree p, Euler's Theorem can be applied. Define the 1-form b by

$$b = A_{ij}(x)x^i dx^j. \tag{10}$$

Now

$$db = \frac{\partial A_{ij}}{\partial x^k}x^i dx^k \wedge dx^j + A_{ij}dx^i \wedge dx^j = \Gamma + A \tag{11}$$

where

$$\begin{aligned}
\Gamma &= \frac{\partial A_{ij}}{\partial x^k}x^i dx^k \wedge dx^j \\[1ex]
&= -\left(\frac{\partial A_{ki}}{\partial x^j} + \frac{\partial A_{jk}}{\partial x^i}\right)x^i dx^k \wedge dx^j \\[1ex]
&= -\frac{\partial A_{ki}}{\partial x^j}x^i dx^k \wedge dx^j + pA \\[1ex]
&= -\Gamma + pA
\end{aligned} \tag{12}$$

Thus

$$A = dB, B = 2b/(p+2).$$

11.   Poisson systems

In the last few years there have been quite a few researchers interested in Poisson systems. See the references in Marsden(1991). Poisson systems are natural generalizations of Hamiltonian systems and there are many interesting physical problems which can be described with this formalism. It turns out that the observation that Lie transforms can be used for tensor fields will be useful in the bifurcation analysis of Poisson systems.

Bifurcation problems are typically local, so a Poisson system will be defined in local coordinates $x^1, \ldots, x^n$. A Poisson structure is a 2-tensor $J = J^{ij}(x)$ which defines a Poisson bracket $\{,\}$ on smooth functions f and g by

$$\{f,g\} = \sum J^{ij}\frac{\partial f}{\partial x_i}\frac{\partial g}{\partial x^j}. \tag{1}$$

A Poisson bracket must satisfy Jacobi's identity. A Poisson system is a system of differential equations of the form

$$\dot{x} = J(x)\frac{\partial H}{\partial x}, \tag{2}$$

where $H = H(x)$ is a smooth function (the Hamiltonian). A Poisson system differs from a classical Hamiltonian system in two respects. First J is not constant and second J may be singular.

There is a generalization of Darboux's Theorem which states that there is a local change of coordinates such that in these coordinates J is constant and of the form

$$J = \begin{pmatrix} 0 & I & 0 \\ -I & 0 & 0 \\ 0 & 0 & 0 \end{pmatrix}.$$

In such coordinates a Poisson system looks like a classical Hamiltonian system with some parameters. The parameters are just the variables which define the null space of J.

If the system depends on a small parameter ϵ then the method of Lie transforms can be used on the tensor J to bring it to the above simple form order by order as discussed in the last section. In most cases the Hamiltonian can be brought into the classical normal form simultaneously. Thus except in some degenerate cases the classical methods and theorems can be adapted to these Poisson systems.

References

Abraham R. and Marsden J. 1978: *Foundations of Mechanics*, Benjamin-Cummings, London.

Birkhoff G. D. 1926: *Dynamical Systems*, Am. Math. Soc., Providence, R.I.

Bruno A. D. 1971: *Analytic forms of differential equations*, Trans. Moscow Math. Soc. 131-228.

Bruno A. D. 1972: *Analytic forms of differential equations II*, Trans. Moscow Math. Soc. 199-239.

Cherry T. M. 1928: *On periodic solutions of Hamiltonian systems of differential equations*, Phil. Trans. Roy. Soc. A 227, 137-221.

Chow S-N. and Hale J. K. 1982: *Methods of Bifurcation Theory*, Springer-Verlag, New York.

Cushman R., Deprit A. and Mosak R. 1983: *Normal forms and representation theory*, J. Math. Phy. 24(8), 2102-2116.

Deprit A. 1969: *Canonical transformation depending on a small parameter*, Celestial Mechanics 72, 173-79.

Diliberto S. P. 1961: *Perturbation theorems for periodic systems*, Circ. Mat. Palermo 9(2), 265-299, 10(2), 111-112.

Diliberto S. P. 1967: *New results on periodic surfaces and the averaging principle*, Differential and Integral Equations, Benjamin, New York, 49-87.

Elphick C., Tirapegui E., Brachet M., Coullet P., Iooss G. 1987: *A simple global characterization for normal forms of singular vector fields*, Physica 29D, 96-127.

Flanders H. 1963: *Differential Forms*, Academic Press, New York.

Golubitsky M. and Schaeffer D. 1986: *Singularities and Groups in Bifurcation Theory I*, Springer-Verlag, New York.

Hartman P. 1964: *Ordinary Differential Equations*, John Wiley, New York.

Henrard J. 1970a: *Periodic orbits emanating from a resonant equilibrium*, Celestial Mechanics 1, 437-66.

Henrard J. 1970b: *On a perturbation theory using Lie transforms*, Celestial Mechanics 3, 107-120.

Henrard J. 1973: *Liapunov's center theorem for resonant equilibrium*, Journal of Differential Equations 14(3), 431-41.

Herman M. R. 1979: *Sur la conjugation différentiable des difféomorphismes du cercle*, I.H.E.S. Math. 49.

Herman M. R. 1983: *Sur les courbes invariantes par les difféomorphismes de l'anneau I*, Astérisque 103, 3-221.

Herman M. R. 1986: *Sur les courbes invariantes par les difféomorphismes de l'anneau II*, Astérisque 144, 3-221.

Hori G. 1966: *Theory of general perturbations with unspecified canonical variables*, Publ. Astron. Soc. Japan 18(4), 287-296.

Kamel A. 1970: *Perturbation method in the theory of nonlinear oscillations*, Celestial Mechanics 3, 90-99.

Kobayashi S. and Nomizu K. 1969: *Foundations of Differential Geometry*, Interscience, New York.

Kummer M. 1976: *On resonant non linearly coupled oscillators with two equal frequencies*, Communications in Mathematical Physics 48, 53-79.

Kummer M. 1978: *On resonant classical Hamiltonians with two equal frequencies*, Communications in Mathematical Physics 58, 85-112.

Marsden J. E. 1991: *Lectures on Mechanics*, Cambridge Univ. Press, Cambridge.

Meyer K. R. 1974: *Normal forms for Hamiltonian systems*, Celestial Mechanics 9, 517-22.

Meyer K. R. 1984a: *Scaling Hamiltonian systems*, SIAM Journal on Mathematical Analysis 15 (5), 877-89.

Meyer K. R. 1984b: *Normal forms for the general equilibrium*, Funkcialaj Ekvacioj 27 (2), 261-71.

Meyer K. R. 1986: *Counter-examples in dynamical systems via normal form theory*, SIAM Review 28 (1), 41-51.

Meyer K. R. 1987: *Bifurcation of a central configuration*, Celestial Mechanics 40 (3), 273-82.

Meyer K. R. 1990: *A Lie transform tutorial II*, Computer Aided Proofs in Analysis (Ed. K. R. Meyer & D. S. Schmidt), IMA Series 28, Springer-Verlag.

Meyer K. R. and Hall, G. R. 1992: *Introduction to Hamiltonian Dynamical Systems*, Springer-Verlag, New York.

Meyer K. R. and Schmidt, D. S. 1971: *Periodic orbits near L_4 for mass ratios near the critical mass ratio of Routh*, Celestial Mechanics 4, 99-109.

Meyer K. R. and Schmidt, D. S. 1977: *Entrainment domains*, Funkcialaj Ekvacioj 20 (2), 171-92.

Meyer K. R. and Schmidt, D. S. 1986: *The stability of the Lagrange triangular point and a theorem of Arnold*, Journal of Differential Equations 62 (2), 222-36.

Meyer K. R. and Schmidt, D. S. 1988: *Bifurcations of relative equilibria in the N-body and Kirchhoff problems*, SIAM Journal on Mathematical Analysis, 19 (6), 1295-1313.

Moser J. K. 1968: *Lectures on Hamiltonian systems*, Mem. Amer. Math. Soc. 81.

Poincaré H. 1885: *Sur les courbes définies par les équations différetielles*, J. Math. Pures Appl. 4, 167-244.

Poinhoff H. G. 1973: *An analytic closing lemma*, Proceedings of the Midwest Dynamical Systems Seminar, Northwestern Univ. Press, Evanston, Ill, 128-256.

Poston T. and Stewart, I. 1978: *Catastrophe Theory and its Applications*, Pitman, Boston.

Schmidt D. S. 1974: *Periodic solutions near a resonant equilibrium of a Hamiltonian system*, Celestial Mechanics 9, 91-103.

Siegel C. L. and Moser, J. K. 1971: *Lectures on Celestial Mechanics*, Springer-Verlag, Berlin.

Sternberg S. 1959: *The structure of local homeomorphisms*, Amer. Math. Soc. 81, 578-604.

Weinstein A. 1971: *Symplectic manifolds and their Lagrangian submanifolds*, Adv. Math. 6, 329-346

Chapter 3

Normal forms

J. Della Dora, L. Stolovitch

Normal forms of differential systems

J. A. Sanders

Versal normal form computation and representation theory

L. Brenig, A. Goriely

Painlevé analysis and normal forms

Y. Sibuya

Normal forms and Stokes multipliers of nonlinear meromorphic differential equations

Normal forms of Differential Systems

Jean Della Dora *   *Laurent Stolovitch* *

To the memory of Jean Martinet,
who highlighted, by his work,
the depth and the beauty of Mathematics

Contents

1. Differential equations   145

2. Linearization of vector fields   147

3. Normal forms of differential systems   151

    3.1. Formal classification of vector fields 153

    3.2. On the effectiveness of normal form computations 161

    3.3. Normal form of systems with a nilpotent linear part 163

4. Normalization of analytic vector fields   164

    4.1. Linearization of holomorphic vector fields 165

    4.2. Normal forms of analytic systems 168

5. Classification of 1-dimensional vector fields   170

    5.1. Formal classification of differential equations on the plane . . . 171

    5.2. Analytic classification of resonant
          diffeomorphisms 173

          5.2.1. Asymptotic expansions and sectorial diffeomorphisms . 174

          5.2.2. Analytic classification of diffeomorphisms formally con-
                  jugate to $g_{1,0}$ 176

*LMC-IMAG, 46 avenue Félix Viallet, 38031 Grenoble Cédex, France

6. **Classification of 2-dimensional vector fields**   **180**

    6.1. Semi-simple case 181

    6.2. Nilpotent case 181

1. Differential equations

In this section, we shall establish some notation and definitions concerning vector fields. Throughout this paper, we shall only be interested in systems of non-linear autonomous differential equations with singularities. For the theory of linear differential systems, with singularities we refer to [Mal91] and [Sib90]. Throughout this section, we shall denote, by E, a K-vectorspace of finite dimension n $(K = R \text{ or } K = C)$.

Definition 1.1 *A* **vector field** *X on an open subset U of E is a map on U into E. A* **trajectory solution** *of X or solution of the differential equation*

$$\dot{x} = X(x), \ x \in U$$

is a parametrized differentiable path $\gamma : I \to U$, where I is an open set of K, such that

$$\forall t \in I, \ \dot{\gamma}(t) = X(\gamma(t))$$

Let X be a vector field on an open set U of E.

Definition 1.2 (Diffeomorphisms) *We shall say that a map $f : U \to V$, where U and V are both open sets of E, is a C^k-diffeomorphism if*

1. *f is of class C^k,*

2. *f is invertible and its inverse is also of class C^k.*

Definition 1.3 (Conjugacy of vector fields) *Let X and Y be two vector fields defined respectively on U and V, two open sets of E. We shall denote, by Φ(resp. Φ'), the flow of X(resp. of Y). We shall denote, by Δ(resp. Δ'), its definition set.*

*We shall say that they are C^k-**conjugate** if there exists a C^k-diffeomorphism $f : V \to U$ such that if $(x, t) \in \Delta$ and $\Phi(x, t) \in U$ then*

1. *$(f(x), t) \in \Delta'$ and $\Phi'(f(x), t) \in U$,*

2. *$\Phi'(f(x), t) = f(\Phi(x, t))$*

If f is differentiable, then the conjugacy relation can be written

$$Df(x)Y(x) = X(f(x))$$

Definition 1.4 *We shall call a* **Lie algebra** *an algebra $\mathcal{G}$ with a non-commutative inner multiplication law $[.,.]$, that is*

$$\forall (x, y) \in (\mathcal{G})^2, \quad [x, y] = -[y, x]$$

which satisfies the Jacobi identity :

$$\forall (x, yz) \in (\mathcal{G})^3, \quad [[x, y], z] + [[y, z], x] + [[z, x], y] = 0$$

This inner multiplication law is called a **Lie bracket**.

For Lie algebra theory, we refer to Postnikov [Pos85] and Varadarajan [Var84].

Example 1.1 Let M be a differentiable manifold of dimension n. Let X be a vector field on M. On a local chart $(U, x_1, \ldots, x_n)$ of the point p, X admits the following representation :

$$X(x) = \sum_{i=1}^{n} X_i(x) \partial_{x_i}$$

where $\{\partial_{x_1}, \ldots, \partial_{x_n}\}$ denote a basis of the tangent space at the point p of M. On the space $\mathcal{X}(M)$ of vector fields on M, we define an inner multiplication law, on a local chart $(U, x_1, \ldots, x_n)$, by

$$[X, Y] = \sum_{i=1}^{n} (\sum_{j=1}^{n} X_j(x) \frac{\partial Y_i}{\partial x_j}(x) - Y_j(x) \frac{\partial X_i}{\partial x_j}(x)) \partial_{x_i}$$

The space $\mathcal{M}(M)$, together with the Lie bracket $[.,.]$, is a Lie algebra.   □

We shall denote, by $\mathcal{X}_n$(resp. $\widehat{\mathcal{X}}_n$), the Lie algebra of holomorphic vector fields in a neighbourhood of the origin of C^n(resp. formal vector fields).

Definition 1.5 *Two maps f, $g : K^n \to K^m$ are* **tangent up to order k** *at a point x, if*

$$\text{in a neighbourhood of } x \quad \|f(y) - \tilde{g}(y)\|_m = o(\|y - x\|_n^k)$$

The relation $\mathcal{R}$ "tangent up to order k at x" is an equivalence relation on the space of functions k-times differentiable at x.

Definition 1.6 *We shall call the* **k-jet** *at the point x of a function f, k-times differentiable at x, the equivalence class of f according to the relation $\mathcal{R}$. This class will be denoted by $j_{k,x}(f)$. A member of this class is the Taylor polynomial of degree k at x. We shall denote, by J_x^k, the* **space of k-jets** *at x. Then, we define natural projections :*

$$j_x^k : C^\infty \to J_x^k$$
$$j_x^{l,k} : J_x^l \to J_x^k \quad \text{if } l \leq k$$

Let $f \in C^\infty(K^n)$, we shall write

$$T_k f(x) = \sum_{l=0}^{k} \frac{D^l f(0)}{l!} \cdot x^{(l)}$$

$$\tilde{T}_k f(x) = \frac{D^k f(0)}{k!} \cdot x^{(k)}$$

where $x^{(l)} \equiv \underbrace{(x, \ldots, x)}_{l \text{ times}}$.

2.   Linearization of vector fields

We shall consider the real autonomous differential system:

$$\frac{dX}{dt} = AX + f(X) \tag{1}$$

where A is an $(n \times n)$-matrix and f a C^r-function in a neighbourhood of 0 such that $f(0) = 0$, $Df(0) = 0$ and $r \geq 2$. Let ϕ_t be its flow.

Then, we consider its "linearized form" at the origin:

$$\frac{dY}{dt} = AY \tag{2}$$

Let ϕ_t^L be its flow.

We wonder to what extent the behaviour of the solutions of the system 1, in a neighbourhood of the fixed point 0, is related to those of the linear system 2. In other words, could the behaviour of the solutions of the non-linear system be deduced from the behaviour of its "linearized form" at the origin. From the mathematical point of view, this question can be represented as follows: Does a local homeomorphism H at the origin, such that $H \circ \phi_t^L = \phi_t \circ H$, exist ? Moreover, it is clear that the more the homeomorphism is regular, the more the solutions of the non-linear system look like those of the "linearized form". So, the next question to be answered is about the regularity of the homeomorphism.

The main result concerning the existence of a homeomorphism which transforms the trajectories of the non-linear system into those of its "linearized form", is known as the Hartman-Grobman theorem. As we shall see, is not always true that such a homomorphism exists.

Theorem 2.1 (Hartman-Grobman theorem) *[PdM82] Let X be a C^1-vector field in a neighbourhood of the origin of R^n. We assume that 0 is a singular point of X.*

*Then if none of the eigenvalues of $DX(0)$ is on the imaginary axis, then
X is topologically conjugate to the linear vector field $x \mapsto DX(0)x$, in a
neighbourhood of 0.*

Example 2.1 Let us consider the following differential system on R^2 :

$$\begin{cases} \dot{x} &= 2x + y^2 \\ \dot{y} &= y + xy \end{cases}$$

The origin is a singular point; the Jacobian matrix of the vector field, at the
origin, is

$$DX(0) = \begin{pmatrix} 2 & 0 \\ 0 & 1 \end{pmatrix}$$

Since none of the eigenvalues is on the imaginary axis, then according to the
Hartman-Grobman theorem, there exists a homeomorphism which sends the
trajectory solutions of the non-linear system to the trajectory solutions of the
system:

$$\begin{cases} \dot{x} &= 2x \\ \dot{y} &= y \end{cases}$$

□

We may wonder what could happen if we do not require the eigenvalues not
to be on the imaginary axis. As the following example will show, the result
is not true any longer.

Example 2.2 Let us consider the following system in R^2:

$$\begin{cases} \dot{x} &= -y - x^3 \\ \dot{y} &= x \end{cases}$$

The origin is a singular point and the linear part, at 0, is $\begin{pmatrix} 0 & -1 \\ 1 & 0 \end{pmatrix}$, where
eigenvalues are $\pm i$. The trajectory solutions of the linear system are circles
centred at the origin. Thus, the origin is stable in the sense of Lyapounov.
Is it the same for the non-linear system? The linearized system is not home-
omorphically conjugate to the non-linear system. In fact, the image, by a
homeomorphism, of a closed trajectory is closed. But, the non-linear system
admits a solution which converges to 0; since such a trajectory is not closed,
this shows that the result is no longer true.   □

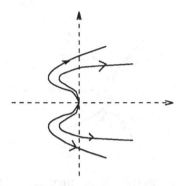

Figure 1. Phase portrait of the non-linear system

We may wonder, if the assumptions of the Hartman-Grobman theorem are fulfilled, about the smoothness of the homeomorphism which linearizes the system. The following example will show that even a simple and very smooth vector field may not be linearized with a smooth function.

Example 2.3 Let us consider the following system on R^2:

$$\begin{cases} \dot{x} &= 2x + y^2 \\ \dot{y} &= y \end{cases} \tag{3}$$

This is a real analytic system on R^2, which is singular at 0; it can be easily integrated and its flow is given by:

$$\forall (x,y) \in R^2, \ \forall t \in R, \quad \Phi_t(x,y) = (e^{2t}(x + y^2 t), \ e^t y)$$

The flow is shown on figure 1.

Let us consider the linearized system of the system (3) at 0:

$$\begin{cases} \dot{x} &= 2x \\ \dot{y} &= y \end{cases} \tag{4}$$

The flow of this linear system is given by:

$$\forall (x,y) \in R^2, \ \forall t \in R, \quad \Phi_t^L(x,y) = (e^{2t}x, \ e^t y)$$

this is shown on figure 2.

According to theorem 2.1, the flows are topologically conjugate to each other; that is, there exist a neighbourhood U of $0 \in R^2$ and a homeomorphism

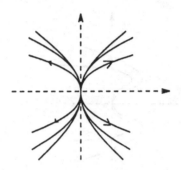

Figure 2. Phases portrait of the linearized of the system

$H : U \rightarrow U$ such that $\forall (x,y) \in R^2$, $\forall t \in R$ such that $\Phi_t^L(x,y)$ is defined and belongs to U, then $\Phi_t(H(x,y))$ exists and belongs to U and we have $\Phi_t(H(x,y)) = H(\Phi_t^L(x,y))$.

We wonder if these systems are C^k-conjugate in a neighbourhood of 0, for some $k \geq 2$.

Proposition 2.1 *The flows of the systems (3) and (4) are not C^2-conjugate in a neighbourhood of the origin.*

Proof :   In fact, let us assume that the flows of the systems 3 and 4 are C^2-conjugate by $H = (h_1, h_2)$ in a neighbourhood U of 0. The equation of conjugacy can thus be written: $\forall (x,y) \in R^2$, $\forall t \in R$ such that $\Phi_t^L(x,y) \in U$:

$$(h_1(x,y) + h_2^2(x,y)t)e^2t = h_1(xe^{2t}, ye^t) \qquad (5)$$
$$h_2(x,y)e^t = h_2(xe^{2t}, ye^t) \qquad (6)$$

Let us differentiate the equation 5 with respect to y; we obtain:

$$(\frac{\partial h_1}{\partial y}(x,y) + 2th_2(x,y)\frac{\partial h_2}{\partial y}(x,y))e^t = \frac{\partial h_1}{\partial y}(e^{2t}x, e^t y)$$

Let us differentiate, once more, with respect to y:

$$(\frac{\partial^2 h_1}{\partial y^2}(x,y) + 2th_2(x,y)\frac{\partial^2 h_2}{\partial y^2}(x,y) + 2t(\frac{\partial h_2}{\partial y}(x,y))^2)e^t = e^t\frac{\partial^2 h_1}{\partial y^2}(e^{2t}x, e^t y)$$

Then we apply the above formula at the point $(x,y) = (0,0)$. Since $h_2(0,0) = 0$, then for all $t \neq 0$, we have:

$$\frac{\partial^2 h_1}{\partial y^2}(0,0) + 2t(\frac{\partial h_2}{\partial y}(0,0))^2 = \frac{\partial^2 h_1}{\partial y^2}(0,0)$$

This fact implies that $\frac{\partial h_2}{\partial y}(0,0) = 0$.

Moreover, let us differentiate the equation (6) with respect to x and let us apply the result at the point $(x,y) = (0,0)$; we obtain:

$$\frac{\partial h_2}{\partial x}(0,0) = e^t \frac{\partial h_2}{\partial x}(0,0)$$

This fact implies that $\frac{\partial h_2}{\partial x}(0,0) = 0$.

Thus, the Jacobian matrix of H at the origin is not invertible; this contradicts the fact that H is a local diffeomorphism at the origin.   □

This phenomenon is due to the fact that, as we shall see in the next section, the non-linear polynomial vector field $y^2 \partial_x$ of the system (3) is a *resonant* term. It cannot be removed by a local C^2-diffeomorphism at the origin.   □

3.   Normal forms of differential systems

Let us consider a holomorphic vector field F, in a neighbourhood of the origin $0 \in C^n$, null at this point. Generally, it is very difficult to give qualitative results about the behaviour of the trajectory of solutions near the singular point.

Given a matrix, it is not obvious now to find out its "behaviour", that is its different invariant subspaces, their dimensions, its kernel, etc... . Nevertheless, if we have the Jordan form of the matrix, then everything becomes clear; most of the knowledge we need about the matrix can be seen in its Jordan form. Moreover, our knowledge becomes complete, if we know the basis with respect to which the matrix is in its Jordan form. Thus, our understanding of the matrix is complete if we have its Jordan form and the change of basis matrix which conjugates it to its Jordan form.

We shall try to define some kind of Jordan form for vector fields. We shall call them **normal forms**. We shall try, by means of change of coordinates, to transform the vector field F into a vector field which is as linear as possible. Nevertheless, as we shall see, some vector fields cannot be linearized; by means of change of variables, we shall not be able to remove some non-linear terms. These are the terms which are called **resonant terms**. They are the key points of the behaviour of the trajectories of the fields F since they make this behaviour quite different from the behaviour of the "linearized form" of F at the origin.

Before defining the normal forms as well as the way to build them up, let us start with a little calculus which will help us understand the problem.

Let us consider the differential system of C^n :

$$\frac{dX}{dt} = F(X) \qquad (7)$$

Let us assume that F is an analytic vector field in a neighbourhood of the origin, which is null at this point. Thus, we can write $F(X) = AX + f(X)$, where A is a complex $(n \times n)$-matrix, f an analytic function in a neighbourhood of 0, such that $f(0) = 0$ and $Df(0) = 0$.

We shall denote, by H_n^i, the space of homogenous polynomials of degree i with n variables; and by P_n^i, the space of polynomials with n variables and of degree lower than or equal to i. Finally, we shall set $\mathcal{H}_n^i \equiv (H_n^i)^n$.

Using the above notation, in a neighbourhood of 0, f can be written:

$$f(X) = f_2(X) + f_3(X) + \cdots + f_p(X) + \cdots$$

where $\forall i \geq 2, \ f_i \in \mathcal{H}_n^i$.

Let us apply the following change of variables to the system (7):

$$Y = H(X) \equiv X + h(X) \qquad (8)$$

where $h \in C[[x_1, \ldots, x_n]]$, $h(0) = 0$ and $Dh(0) = 0$.

This change of variables transforms the system 7 into a system of the form

$$\dot{Y} = AY + g(Y) \qquad (9)$$

Thus, we obtain the following relations:

$$\begin{aligned}
\dot{Y} &= \dot{X} + Dh(X)\dot{X} \\
&= (Id + Dh(X))(AX + f(X)) \\
&= AX + Ah(X) + g(X + h(X))
\end{aligned}$$

which leads to the following functional equation:

$$Dh(X)(AX) - A(h(X)) = g(X + h(X)) - f(X) - Dh(X)f(X) \qquad (10)$$

Thus, the systems (7) and (9) are locally equivalent at 0, by the diffeomorphism H defined by (8), if and only if the equation (10) is satisfied. Let us consider the linear oprator:

$$\begin{aligned}
L_A: \quad & G_0 \quad \rightarrow \quad G_0 \\
& h(X) \quad \mapsto \quad Dh(X)(AX) - A(h(X))
\end{aligned}$$

where G_0 is a space of formal maps of C^n, null at 0 and with null derivative at 0. The equation (10) can thus be written

$$L_A(h)(X) = g(X + h(X)) - f(X) - Dh(X)f(X)$$

It is clear that the spaces $\mathcal{H}_n^i$ are stable under the action of L_A. By doing a Taylor expansion to order 2 of the equation (10), we obtain

$$L_A(h_2)(X) = g_2(X) - f_2(X)$$

If the restriction of L_A to $\mathcal{H}_n^2$ is invertible, then we can set $g_2 = 0$, whereas h_2 is uniquely determined. In this case, we have linearized the initial system up to order 3, since the system (9) does not have any non-linear terms of order 2.

Otherwise, we split the space $\mathcal{H}_n^2$ in the following way: $\mathcal{H}_n^2 = Im(L_A) \oplus S_2$. Thus, we can find h_2 such that $L_A(h_2) = \pi_{Im(L_A)}(-f_2)$. In this case, in order to solve the equation, we shall have to set $g_2 = \pi_{S_2}(f_2)$; this term is the non-linear term of order 2 of the normal form of the system (7). Thus, some non-linear terms of order 2 will remain: these terms are called **resonant**.

We shall have to characterize the fact that the restriction of L_A to $\mathcal{H}_n^2$ is invertible, the space S_2 as well as the means to compute the normal form up to order 3. Then, we shall have to solve the problem up to order $k \geq 2$, the problem being solved up to order $k - 1$.

One could consult [Arn80], [Bru89] and [Van]. For applications of the normal form theory, we refer to [Sta85] and [Bru89].

3.1.   Formal classification of vector fields

In this section, we shall be concerned with formal vector fields and their classification by "formal diffeomorphisms".

Let F be a formal vector field on C^n. We assume that it is singular at the origin, i.e. $F(0) = 0$. Even if this means applying a linear change of coordinates, we may assume that the Jacobian matrix $DF(0)$ of F at 0 is a Jordan form.

Let Φ be a "formal diffeomorphism" of C^n fixing the origin, that is an n-dimensional vector whose coordinates are formal power series null at 0 such that $D\Phi(0)$ is invertible. Moreover, we shall assume that Φ is "tangent to identity" at 0, i.e $D\Phi(0) = Id$.

Let us consider the action of Φ on F: $G \equiv \Phi * F$. By definition, we have:

$$D\Phi(x)G(x) = F(\Phi(x)) \tag{11}$$

Lemma 3.1 *Let G be the vector field defined above, then:*

1. $G(0) = 0$,

2. $DG(0) = DF(0)$.

Thus, we shall write $G(X) = DF(0)X + g(X)$, with $Dg(0) = 0$ and $g(0)$. We shall need a formula which expresses the different derivatives of g:

Lemma 3.2 *For all integer $k \geq 2$, we have*

$$\tilde{T}_k g = (ad_k A)(\tilde{T}_k \Phi) + \tilde{T}_k[(T_k f - a) \circ (T_{k-1}\Phi) - L_{T_{k-1}g - A}(T_{k-1}\Phi - I)]$$

(we recall that $T_k g$ denotes the Taylor polynomial of order k of g whereas $\tilde{T}_k g$ denotes the homogeneous polynomial of degree k of Taylor expansion of g, and that $L_A(h)(x) = Dh(X)(AX) - A(h(X)))$.
Proof :   In fact, 11 can be written in the following way:

$$\begin{aligned} g &= f \circ \Phi - L_g(\Phi - I) \\ &= A \circ \Phi - L_A(\Phi - I) + (f - A) \circ \Phi - L_{g-A}(\Phi - I) \end{aligned}$$

By applying $\tilde{T}_k$, for all $k \geq 2$, we obtain

$$\tilde{T}_k g = (ad_k A)(\tilde{T}_k \Phi) + \tilde{T}_k[(f - a) \circ \Phi - L_{g-A}(\Phi - I)]$$

But, $T_1(f - A) = T_1(g - A) = 0$, thus

$$\tilde{T}_k[(f-a)\circ\Phi - L_{g-A}(\Phi - I)] = \tilde{T}_k[(T_k f - a) \circ (T_{k-1}\Phi) - L_{T_{k-1}g - A}(T_{k-1}\Phi - I)]$$

which proves the lemma.   □

We shall state and prove the main result about the normalization of vector fields: the Poincaré-Dulac theorem. Before doing so, let us recall a classical result about endomorphisms.

Theorem 3.3 ($S + N$ decomposition) *Let A be an endomorphism of a finite-dimensional R-vectorspace E. Then, there exist two endomorphisms S and N of E such that:*

1. $A = S + N$,

2. $[S, N] \equiv S \circ N - N \circ S = 0$,

3. S is semi-simple,

4. F is nilpotent.

If A is a Jordan form, then the decomposition is obvious.

Definition 3.1 Let F be a formal vector field which is null at the origin. We shall say that F is a **nilpotent** vector field if its Jacobian matrix at the origin $DF(0)$ is nilpotent.

Theorem 3.4 (Poincaré-Dulac theorem) Let $F \in \widehat{\mathcal{X}_n}$ be a formal vector field whose 1-jet at the origin is the matrix A, which is assumed to be a **Jordan form**. Let S and N be the matrices of the $S + N$ decomposition of A.

Then there exists a formal diffeomorphism Φ which is tangent to identity at 0 such that

$$\Phi * F = S + N'$$

where :

1. S designates a semi-simple linear vector field $x \rightarrow Sx$,

2. N' is a nilpotent vector field with $DN'(0) = N$,

3. $[S, N'] = 0$, $[.,.]$ designates the Lie bracket of vector fields.

The vector field $S + N'$ is then called the **normal form** of F .

Let A be an $(n \times n)$-matrix which is a Jordan form. Let $S + N$ be its Dunford decomposition. In the next chapter, we shall denote by A either the matrix A or the vector field $x \rightarrow Ax$.

Let us consider the adjoint application $ad\, S$ of S on the Lie algebra of formal vector fields $\widehat{\mathcal{X}_n}$ which are null at the origin. We shall denote by $\lambda_1, \ldots, \lambda_n$ the eigenvalues of A. Since S is a diagonal matrix, the vector field S can be written

$$S = \sum_{i=1}^{n} \lambda_i x_i \partial_i$$

Let us compute the adjoint of S:

$$[S, Y] = \sum_{i=1}^{n} (\sum_{j=1}^{n} \lambda_j x_j \frac{Y_i}{\partial x_j} - Y_j \frac{\lambda_i x_i}{\partial x_j}) \partial_i$$

that is

$$[S, Y] = \sum_{i=1}^{n} ((S.Y_i) - \lambda_i Y_i) \partial_i$$

Consequently, Y is an eigenvector of $ad\, S$ if and only if there exists $\mu \in C$ such that

$$\forall 1 \leq i \leq n, \quad (S.Y_i) - \lambda_i Y_i = \mu Y_i$$

But, for all n-tuples of integers $Q = (q_1, \ldots, q_n) \in N^n$, we have

$$S.x^Q = (\sum_{i=1}^{n} q_i \lambda_i) x^Q = (Q, \lambda) x^Q$$

where we have set $\lambda = (\lambda_1, \ldots, \lambda_n)$ and $(Q, \lambda) \equiv \sum_{i=1}^{n} q_i \lambda_i$.

We deduce, from that fact, that $ad\, S$ admits the vector field $x^Q \partial_s$ as eigenvector associated to the eigenvalue $(Q, \lambda) - \lambda_s$. Thus, we have proved the following result:

Proposition 3.1 *Let S be a diagonal $(n \times n)$-matrix; let $\lambda_1, \ldots, \lambda_n$ be its eigenvalues.*

Then, the operator $ad\, S$ on $\widehat{\mathcal{X}_n}$ satisfies the following properties:

1. *the set $sp(S)$ of its eigenvalues is composed of the numbers $(Q, \lambda) - \lambda_s$, with $Q \in N^n$ and $1 \leq s \leq n$,*

2. *the eigenspace E_a of $ad\, S$, associated to the eigenvalue a, is generated by the vector fields of the form $x^Q \partial_s$ such that $(Q, \lambda) - \lambda_s = a$.*

The subspace E_0 is particuliary interesting since it is composed of vector fields which commute with S, that is $\forall Y \in E_0$, $[S, Y] = 0$. Moreover, a monomial vector field $x^Q \partial_s$ belongs to E_0 if and only if $(Q, \lambda) = \lambda_s$. Nevertheless, the non-linear terms which interest us are the ones which commute with S, which is why we are led to state the following definition.

Definition 3.2 *Let $\lambda_1, \ldots, \lambda_n$ be n complex numbers. Let $k \geq 2$ be an integer.*

*We shall say that these numbers satisfy a **resonance relation of order** k, if there exists*

1. *an n-tuple of integers Q such that $|Q| \equiv \sum_{i=1}^{n} q_i = k$,*

2. *an index $1 \leq s \leq n$,*

such that

$$(Q, \lambda) \equiv \sum_{i=1}^{n} q_i \lambda_i = \lambda_s.$$

To such a resonance relation, one associates the **resonant monomial vector field** $x^Q \partial_s$.

Let E be the subspace of $\widehat{\mathcal{X}_n}$, which is the direct sum of the subspaces E_a, $a \in sp(S)$, $a \neq 0$. The following proposition is the key point in the decomposition of the space $\mathcal{H}_n^i$.

Proposition 3.2 *Let A be an $(n \times n)$-Jordan matrix. Let $A = S + N$ be its Dunford decomposition. As previously mentioned, E_0 will denote the kernel of $ad\, S$ and E the sum of the other eigenspaces of $ad\, S$ associated to the non-zero eigenvalues.*

Then, for all vector fields $Y \in \widehat{\mathcal{X}_n}$, there exist two vector fields, uniquely determined, $(Y_0, Z) \in (\widehat{\mathcal{X}_n})^2$, such that

  1. $Y_0 \in E_0$ and $Z \in E$,

  2. $Y = Y_0 + [A, Z]$.

Moreover, if Y is k-flat, i.e. $D^\sigma Y(0) = 0$ for all $|\sigma| \leq k$, then Y_0 and Z are both k-flat.

Proof :   In fact, we have the splitting $\widehat{\mathcal{X}_n} = E_0 \oplus E$; consequently, $Y \in \widehat{\mathcal{X}_n}$ is the sum of two elements $Y_0 \in E_0$ and $Y_1 \in E$, both uniquely determined. But, the matrix N commutes with S; thus, the stable subspaces under S are stable under N, and conversely. Moreover, $ad\, A$ is invertible on E; thus there exists a unique Z such that $(ad\, A)Z = Y_1$.   □

Proof of theorem 3.4:   Let us prove, by induction on m, that

$$\forall 2 \leq k \leq m, \quad \tilde{T}_m[\Phi * F] \in E_0$$

Let us assume that the proposition is true up to order m.

We set $E_0^m = E_0 \cap \mathcal{H}_n^m$ and $E^m = E \cap \mathcal{H}_n^m$. According to the previous proposition, we can write

$$\forall m \geq 2, \quad \mathcal{H}_n^m = E_0^m \oplus (ad_m A)(E^m)$$

More precisely, we have

$$E_0^m = \{x^Q \partial_s \mid (Q, \lambda) = \lambda_s, \, |Q| = m, \, 1 \leq s \leq n \}$$

Then, let Φ be a formal diffeomorphism such that

$$T_m \Phi = Id + \sum_{i=1}^{m} \phi_i \text{ where } \phi_i \in \mathcal{H}_n^i$$

and

$$T_m g = A + \sum_{i=1}^{m} w_i \text{ where } w_i \in E_0^i$$

By means of lemma 3.2, we obtain

$$\tilde{T}_{m+1} g = (ad_{m+1} A)(\tilde{T}_{m+1} \Phi) + \tilde{T}_{m+1}[(T_{m+1} f - a) \circ (T_m \Phi) - L_{T_m g - A}(T_m \Phi - I)]$$

According to what was previously said, there exists $\tilde{T}_{m+1} \Phi \in E^{m+1}$ such that

$$(ad_{m+1} A)(\tilde{T}_{m+1} \Phi) = -\pi_{E^{m+1}}(\tilde{T}_{m+1}[(T_{m+1} f - a) \circ (T_m \Phi) - L_{T_m g - A}(T_m \Phi - I)])$$

where π_E denotes the projection on the subspace E. Then we have

$$\tilde{T}_{m+1} g = \pi_{E_0^{m+1}}(\tilde{T}_{m+1}[(T_{m+1} f - a)(T_m \Phi) - L_{T_m g - A}(T_m \Phi - I)])$$

that is $\tilde{T}_{m+1} g \in E_0^{m+1}$; this ends the induction.

Thus, we have shown that there exists a formal diffeomorphism Φ such that $\Phi * F = Ax + N'$, where the vector field N' is nilpotent, since it is non-linear; moreover, it commutes with the vector field S since it belongs to E_0.

On the other hand, the matrices S and N commute with each other; thus, $[Sx, Nx] = 0$ in $\widehat{\mathcal{X}}_n$, and Nx is a nilpotent vector field. Consequently,

$$\Phi * F = S + \underbrace{N + \tilde{T}_2[\Phi * F] + \cdots}_{N'}$$

□

In a more "concrete" way, the Poincaré-Dulac theorem can be stated as follows:

Theorem 3.5 *Let F be a formal vector field on C^n. We shall assume that F is zero at the origin. Let A be the Jacobian matrix of F at 0. We shall denote by J its Jordan form and by $\lambda_1, \ldots, \lambda_n$ its eigenvalues. The differential equation which is associated to F is*

$$\frac{dx}{dt} = F(x) = Ax + f(x), \quad x \in C^n \tag{12}$$

Then there exists a change of variables $x = \Phi(y)$ which transforms system 12 into the system

$$\frac{dy}{dt} = D\Phi^{-1}(y)F(\Phi(y)) = Jy + \Big(\sum_{\substack{(Q,\lambda)=\lambda_i \\ Q \in N^n, |Q| \geq 2}} a_Q^i y^Q \partial_i \Big)_{1 \leq i \leq n}$$

From the above theorem, we deduce a very interesting result:

Corollary 3.6 (Poincaré theorem) *Let F be a formal vector field on C^n. We asssume that F is zero at the origin. Let A be the Jacobian matrix of F at 0. We shall denote by J its Jordan form, and by $\lambda_1, \ldots, \lambda_n$ its eigenvalues. The differential equation which is associated to F is*

$$\frac{dx}{dt} = F(x) = Ax + f(x), \quad x \in C^n \tag{13}$$

We assume that there is no resonance relation which is satisfied, that is,

$$\forall k \geq 2,\ \forall Q \in N^n \text{ such that } |Q| = k,\ \forall 1 \leq s \leq n,\quad \lambda_s \neq (Q, \lambda)$$

Then there exists a change of variables $x = \Phi(y)$ which transforms system (13) into the system

$$\frac{dy}{dt} = Jy$$

Let us illustrate these results by means of some examples.

Example 3.1 Let us consider the following system on C^n:

$$\frac{dx}{dt} = x + f(x), \quad x \in C^n$$

where f is a formal power series whose terms of lowest degree are, at least, quadratic, that is $f(0) = 0$ and $Df(0) = 0$.

Then there exists a change of variables which transforms the previous system
into the system

$$\frac{dy}{dt} = y$$

In fact, the eigenvalues of the linear part all equal 1; thus, there can be no
resonance relation satisfied.   □

Example 3.2 Let us consider the following system on C^2:

$$\begin{cases} \dot{x} &=& 2x + y^2 + x^2 y \\ \dot{y} &=& y + xy + x^5 \end{cases}$$

Then there exists a change of variables which transforms the previous system
into the system

$$\begin{cases} \dot{x} &=& 2x + y^2 \\ \dot{y} &=& y \end{cases}$$

In fact, the eigenvalues of the linear part are $\lambda_1 = 2$ and $\lambda_2 = 1$. But, the
only resonance relation satisfied is

$$0.\lambda_1 + 2.\lambda_2 = 2 = \lambda_1$$

The associated resonant vector field is $(y^2, 0)$; this explains the result obtained.   □

We shall give another proof of the Poincaré-Dulac theorem by the so-called
"Newton method". This method will be useful, as we shall see in the next
chapter.

Proof of theorem 3.4 by the Newton method: Following Martinet
[Mar80], let us assume the vector field X is normalized up to order k, that is
$X = S + N + R$, where N is a nilpotent polynomial vector field of degree k,
commuting with S, and R a k-flat vector field.

According to proposition 3.2, there exist $R_0 \in E_0$ and $U \in E$, both k-flat,
such that

$$R = R_0 + [S + N, U]$$

We set $\phi = \exp U$. Then,

$$\phi * X = S + N + R_0 + R'$$

where R' is $2k$-flat. In fact, it is well known that

$$\exp tU * X = X + t[U, X] + \frac{1}{2}t^2[U, [U, X]] + \cdots$$

the coefficient of t^p is a pk-flat vector field. Thus, we have

$$\exp U * X = S + N + R_0 + [S + N, U] + [U, S + N] + \ 2k\text{-flat terms}$$

that is,

$$\phi * X = S + N' + R'$$

with $[S, N'] = 0$ and R' a $2k$-flat vector field. Thus, we have normalized X up to order $2k$; that is why this method is called the "Newton method". Moreover, the diffeomorphism $\exp U$ is k-flat with respect to the identity. $\square$

3.2. On the effectiveness of normal form computations

Although the proof of the Poincaré-Dulac theorem we have given seems to be effective, some difficulties remain and are often underestimated in the literature, the problem of resonances being the main one.

Let X be a given vector field on C^n, singular at the origin. Let A be its Jacobian matrix at 0. Let us denote by $\lambda_1, \ldots, \lambda_n$ its eigenvalues. Nevertheless, these numbers are usually unknown. So, the question is how to know if there is a resonance relation of order k? The next question is, which relation is satisfied, that is how to find all the n-tuples $(q_1, \ldots, q_n)$ of integers and all the indices $1 \leq s \leq n$ for which we have

$$\sum_{i=1}^n q_i \lambda_i = \lambda_s \ ?$$

One could think of using numerical approximations of the eigenvalues to solve this problem. Nevertheless, the approximating numbers could satisfy some resonance relations not satisfied by the eigenvalues, and conversely. For instance, if A is a (2×2)-matrix with $e = \exp 1$ and -1 as eigenvalues. No resonance relation is possible. This is not the case for the approximating numbers $a_N = \sum_{k=0}^N \frac{1}{k!}$ and -1. In fact, there exist two integers p and q such that $pa_N + q(-1) = 0$, with $p + q \geq 2$.

Generally, the coefficients of A are rational numbers; thus, its characteristic polynomial has rational coefficients. The **resonance problem** can be stated as follows:

Let $\lambda_1, \ldots, \lambda_n$ be n algebraic numbers given by the polynomial P, with rational coefficients, that they are roots of. Does there exist an algorithm which provides an answer to the question: "For any $k \geq 2$, is there a resonance relation of order k?", exist? Does an algorithm which provides, for all $k \geq 2$, all $(n+1)$-tuples $(q_1, \ldots, q_n, s)$, with $q_1 + \cdots + q_n = k$ and $1 \leq s \leq s$, such that

$$\sum_{i=1}^{n} q_i \lambda_i = \lambda_s \quad ?$$

This problem seems about to be solved, using algebraic number theory methods. Nevertheless, the problem of having a "finite time" algorithm which says whether any resonance relation of any order is satisfied or not, seems to be hopeless.

As we have seen, the linear part of the normal form is the Jordan matrix of the Jacobian matrix, at the origin, of the vector field. Generally, the Jacobian matrix is not a Jordan form; thus, it has to be computed. This computation is far from being trivial. A very efficient algorithm has been developed by Gil [Gil, GV].

After having seen the resonance problem, we have to deal with another problem which is not that easy to solve if we keep in mind the effectiveness. In fact, from a purely theoretical point of view, the normal form problem reduces to the inversion of a linear operator in different finite dimensional vector spaces. Nevertheless, from the effectiveness point of view, some problems arise. In fact, let us assume that we have found a polynomial diffeomorphism ϕ of order k which transforms the vector field X to its normal form up to order k. Then we have to take into account the perturbations on the non-linear terms of order greater than k brought about by the action of ϕ on X. A complete algorithm, following the proof we have shown, has been developed, studied from the complexity point of view and implemented on Axiom, by Vallier [Val].

Another point of view has been adopted by Della Dora and Stolovitch[SDD91], for these computations. It is based on the matrix representation of restrictions, to the k-jet spaces, of a differentiation and an automorphism of the ring of formal power series with n variables and complex coefficients. They have stated a Poincaré-Dulac theorem for differentiations of the ring of formal power series in n **non-commutative** variables, that is a classification theorem of these differentiations by means of the action of an automorphism of the same ring. The proof is effective and provides an algorithm to compute normal forms without having any problem with the computation of the "induced non-linear terms" since everything is given by a matricial induction equation expressed in terms of tensor products of matrices already known and stored.

3.3. Normal form of systems with a nilpotent linear part

The Poincaré-Dulac theorem states that there exists a formal diffeomorphism ϕ such that $\phi * X = S + N$, where S is the diagonal linear vector field associated to the diagonal matrix of the Jordan form of the Jacobian matrix of X at 0, N is nilpotent and commutes with S. When, the Jacobian matrix is nilpotent, then $S = 0$, and one could set $\phi = Identity$. This does not provide any simplification of the vector field! Nevertheless, as we shall show, some simplifications are possible. To our knowledge, there are two different approaches to the problem. One is due to Cushman and Sanders [CS86] and Bogaevskiy, Povzner and Givental [AA88][p 68-71]. The second is due to Stolovitch [Sto92].

In the first approach, a supplementary subspace of the image of $ad(A)$ in the space of homogenous vector fields of degree k is found. This subspace is determined by the means of sl_2 representations. We refer to the paper of Sanders in this volume.

In the second approach, we use the matrix representation of restrictions of a differentiation on the k-jet. By analysing the terms which can be removed, we obtain a normal form and an algorithm which computes it as well as the normalizing diffeomorphism.

Proposition 3.3 *Let X be a formal vector field on C^n, singular at 0, whose linear part is a Jordan block.*

Then there exists a formal diffeomorphism ϕ, such that

$$\phi * X = \sum_{i=1}^{n} \left(\epsilon_i x_{i+1} + \sum_{r \geq 2} \left[\sum_{k=1}^{n-1} \sum_{l=2}^{r} \left[\sum_{\sigma_1 + \cdots + \sigma_{k-1} + l = r} a_{r,k,l,\sigma,i} \, x_1^{\sigma_1} \ldots x_{k-1}^{\sigma_{k-1}} x_k^l \right] \right] \right) \partial_{x_i}$$

where $a_{r,k,l,\sigma,i} \in C$, $\epsilon_i = 1$ if $i < n$ and $\epsilon_n = 0$.

In the case where there is more than one Jordan block, the result is more complicated. We refer to [Sto92].

The result stated above is "visually" complicated. Let us illustrate it on 2-dimensional systems.

Example 3.3 Let us consider the 2-dimensional system

$$\begin{pmatrix} \dot{x} \\ \dot{y} \end{pmatrix} = \begin{pmatrix} 0 & 1 \\ 0 & 0 \end{pmatrix} \begin{pmatrix} x \\ y \end{pmatrix} + \begin{pmatrix} f(x,y) \\ g(x,y) \end{pmatrix}$$

where f and g are formal power series such that $Df(0,0) = Dg(0,0) = 0$ and $f(0,0) = g(0,0) = 0$.

According to the proposition above, there exists a formal diffeomorphism ϕ such that the change of variables $(X,Y) = \phi(x,y)$ transforms the previous system into

$$\begin{pmatrix} \dot{X} \\ \dot{Y} \end{pmatrix} = \begin{pmatrix} 0 & 1 \\ 0 & 0 \end{pmatrix} \begin{pmatrix} X \\ Y \end{pmatrix} + \begin{pmatrix} F(X) \\ G(X) \end{pmatrix}$$

where F and G are formal power series such that $DF(0) = DG(0) = 0$ and $F(0) = G(0) = 0$. This 2-dimemsional particular case can be found in [Tak73]. □

4.   Normalization of analytic vector fields

We shall consider, in this section, holomorphic vector fields in a neighbourhood of a singular point on C^n(this point will always be the origin). In order to have a more complete understanding of the behaviour of the trajectory solutions, in a neighbourhood of the origin, we shall perform a normalization of the vector field, that is, we shall transform it into a normal form, by means of a change of variables.

This change of variables ϕ is an n-dimensional vector whose components are formal power series. If we want to deduce qualitative results about the vector field from qualitative results obtained on its normal form, a question arises:

Is ϕ analytic in some neighbourhood of the origin?

As we shall see, this is not always the case. So, the next question to be asked is whether these formal power series can be regarded as asymptotic expansions, in some sectors of the origin, of analytic functions which normalize the vector field?

The last problem has been solved by Ramis and Martinet for 2-dimensional vector fields [RM82, RM83]. These results involve the theory of summation of formal power series due to Ramis [Ram93].

For the linearization problem in the C^∞ and C^k cases, we refer to [AA88], [Sel85], [Bel78] and [Ste58, Ste59].

4.1.   Linearization of holomorphic vector fields

We shall be interested in the linearization of holomorphic vector fields on C^n. As we have shown, if the eigenvalues of the linear part of a vector field X are not resonant, then X is formally linearizable, that is there exists a formal diffeomorphism ϕ such that $\phi * X(x) = DX(0)x$. We wonder to what extent the diffeomorphism ϕ is analytic in a neighbourhood of the origin.

Unfortunately, as we shall see, the linearizing transformation can be divergent. Roughly, the numbers $\frac{1}{(Q,\lambda)-\lambda_s}$, which exist, since the eigenvalues are non-resonant, intervene in the coefficients of the Taylor expansion of the linearizing diffeomorphism. Although non-zero, the numbers $(Q,\lambda) - \lambda_s$ can be sufficiently small to make the linearizing series divergent; this phenomenon is known as the **small divisors** problem.

The first answer, about analytic linearization, was given by Siegel in 1952. By requiring the small divisors not to be too small, he proved that the linearizing transformations are analytic.

Theorem 4.1 (Siegel theorem) *[Ste69] Let X be an analytic vector field in a neighbourhood of 0 in C^n. We assume that the origin is a singular point, i.e. $X(0) = 0$. Moreover, we assume that*

1. *the eigenvalues $\lambda_1, \ldots, \lambda_n$ of $DX(0)$ do not satisfy any resonance relation,*

2. *there exist $C > 0$ and $\nu > 0$ such that*

$$\forall k \geq 2, \forall 1 \leq s \leq n, \forall Q \in N^n \text{ such that } |Q| = k, \quad |(Q,\lambda) - \lambda_s| \geq \frac{C}{|Q|^\nu}$$

Then F is holomorphically conjugate to its linear part, in a neighbourhood of 0.

The result above depends on an arithmetical condition. This condition has been improved by Bruno.

Theorem 4.2 (Bruno theorem) *[Bru89, Bru72] Let X be a holomorphic vector field in a neighbourhood of the origin, null at this point. Let $\lambda_1, \ldots, \lambda_n$ be the eigenvalues of $DX(0)$, assumed not to be resonant. We set, for all $k \geq 0$,*

$$\omega_k = \min\{ |(Q,\lambda) - \lambda_s| \mid |Q| \leq 2^{k+1}, 1 \leq s \leq n\}$$

If the power series

$$\sum_{k=0}^{\infty} \frac{\log \frac{1}{\omega_k}}{2^k} \tag{14}$$

converges, then X is holomorphically conjugate to its linear part, in a neighbourhood of 0.

We shall follow Martinet [Mar80] and prove the theorem in the case where the linear part S is diagonal. First of all we fix some notation. Let $\rho > 0$ be a real number. We set

$$D_r = \{x \in C^n | \, |x_i| \leq r\}$$

Let $f = \sum_{Q \in N^n} a_Q x^Q$ be an analytic function on D_r; we set

$$\|f\|_r = \sum_{Q \in N^n} |a_Q| r^{|Q|}.$$

As usual, in $Q = (q_1, \ldots, q_n) \in N^n$, we shall write

$$|Q| = \sum_{k=1}^{n} q_k \text{ and } x^Q = x_1^{q_1} \ldots x_n^{q_n}$$

If X denotes a vector field on D_r, then we shall denote, by $\|X\|_r$ the greatest norm of its components. Let $\lambda_1, \ldots, \lambda_n$ be the eigenvalues of $DX(0)$. There is no need to assume that $\omega_k \leq 1$. We set

$$\sigma_k = \frac{\omega_k^{\frac{1}{m}}}{m^{\frac{2}{m}}} \text{ and } \tau_k = \frac{\omega_k^{\frac{1}{m}}}{m^{\frac{1}{m}}}$$

where $m = 2^k$. We have $\sigma_k < \tau_k < 1$ and $\lim_{k \to +\infty} \sigma_k = 1$, according to the convergence of the series 14.

We shall normalize the vector field according to the Newton method shown above. Nevertheless, we shall be concerned with the "growth" and the convergence of the solution. This is answered by the following lemma, which is the key point of the convergence of the construction.

Lemma 4.3 *Let $\frac{1}{2} < r < 1$ be a real number. Let Y be an analytic vector field on D_r which is 2^k-flat at the origin and such that $\|Y\|_r < 1$. According to proposition 3.2, we have the decomposition*

$$Y = [S, Z] \text{ with } Z \in E \text{ being } 2^k\text{-flat}$$

since there is no resonance relation.

Let U be the polynomial vector field component of Z of degree 2^{k+1}, then

1. $\|U\|_\rho < \frac{1}{2^k}$, where $\rho = \tau_k r$,

2. $D_{\rho_1} \subset \phi(D_\rho)$ where $\phi = \exp U$ and $\rho_1 = \sigma_k r$,

3. $\|Y_1\|_{\rho_1} < 1$ where $\phi * (S + Y) = S + Y_1$.

Proof :   From the properties of $ad\,S$, we have :

if $Y = (\sum_{Q \in N^n, |Q| > 2^k} a_{Q,i} x^Q)_{1 \le i \le n}$, then the solution Z of $ad\,S(Z) = Y$ is

$$Z = (\sum_{|Q| > 2^k} \frac{a_{Q,i}}{(Q, \lambda) - \lambda_i} x^Q)_{1 \le i \le n}$$

Thus, by the definition of ω_k, we have

$$\begin{aligned} \|U\|_r &= \sup_{1 \le i \le n} \sum_{2^{k+1} \ge |Q| > 2^k} \frac{|a_{Q,i}|}{|(Q, \lambda) - \lambda_i|} r^{|Q|} \\ &\le \frac{1}{\omega_k} \sup_{1 \le i \le n} \sum_{2^{k+1} \ge |Q| > 2^k} |a_{Q,i}| r^{|Q|} \\ &\le \frac{1}{\omega_k} \|Y\|_r < \frac{1}{\omega_k} \end{aligned}$$

which gives the first point.

Moreover, since U is a 2^k-flat vector field, we have

$$\|U\|_\rho \le (\frac{\rho}{r})^m \|U\|_r < \frac{1}{2^k}$$

But, it is easy to check that, for sufficiently large k, $\sigma_k + \frac{1}{r 2^k} < \tau_k$. Thus, since $\tau_k r - \frac{1}{2^k} \le \|Id\|_\rho - \|U\|_\rho \le \|\exp U\|_\rho$, then we have $D_{\rho_1} \subset \phi(D_\rho)$. $\square$

Proof of theorem 4.2:

Let X be a holomorphic vector field, singular at the origin and with non-resonant eigenvalues (of $DX(0)$). They are assumed to satisfy the arithmetical condition of convergence of the series (14). According to this condition, the infinite product $\prod \sigma_k$ converges; thus, we can choose an integer p such that

$$\frac{1}{2} < \prod_{k \ge p} \sigma_k < 1$$

Let us assume that X is linearized up to order 2^p, that is, $X = S + Y_p$, where Y_p is a 2^p-flat vector field. Moreover, we can assume, without loss of generality, that Y_p is holomorphic on D_1 and $\|Y_p\|_1 < 1$. According to the Newton method of normalization together with the previous lemma applied to $Y = Y_p$

and $r = 1$, there exists a diffeomorphisms ϕ_p such that $\phi_p * X = S + Y_{p+1}$ where Y_{p+1} is a 2^{k+1}-flat vector field. From the lemma, we have that Y_{p+1} is analytic on D_{σ_p} and $\|Y_{p+1}\|_{\sigma_p} < 1$. Moreover, the diffeomorphism ϕ_p is $\frac{1}{2^p}$-flat with respect to the identity. Consequently, the sequence of diffeomorphisms $(\phi_k \circ \cdots \circ \phi_p)_{k \geq p}$ converges to an analytic diffeomorphism ϕ defined on the disk D_σ where $\sigma = \prod_{k \geq p} \sigma_k$, and satisfies $\phi * X = S$.   □An arithmetical condition on the eigenvalues is necessary. In fact, we have the following result.

Theorem 4.4 *With the notation of theorem 4.2, if the eigenvalues satisfy the arithmetical condition*

$$\limsup_{k} \frac{\ln \frac{1}{\omega_k}}{2^k} = +\infty \tag{15}$$

then there exists a non-linear holomorphic vector field X whose linearizing transformation diverges.

We have one result on the convergence of the linearizing transformations, related to the arithmetical condition 14, and one result on the divergence, related to the arithmetical condition 15. Nevertheless, these conditions are not contrary to each other. So, the natural question which arises is what are the analytic proprieties of linearizing transformations when

$$\sum_{k=0}^{\infty} \frac{\log \frac{1}{\omega_k}}{2^k} \text{ diverges, and } \limsup_{k} \frac{\ln \frac{1}{\omega_k}}{2^k} < +\infty$$

This problem has been analysed by Yoccoz [Yoc85] in the case of diffeomorphisms of the complex plane.

4.2. Normal forms of analytic systems

We wonder what happens if the resonance relations are satisfied; in this case, the vector field is not linearizable. We shall start with the simplest case.

Definition 4.1 *We shall say that n complex numbers $\lambda_1, \ldots, \lambda_n$ belong to the Poincaré domain if their convex hull $\mathcal{E}(\lambda_1, \ldots, \lambda_n)$ does not contain the origin; otherwise, we shall say that they belong to the Siegel domain. See figures 3 and 4*

Proposition 4.1 *Let $\lambda_1, \ldots, \lambda_n$ be n complex numbers belonging to the Poincaré domain.*
Then there exists only a finite number of resonance relations.

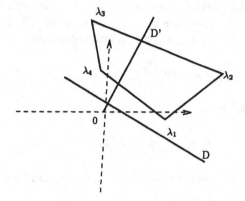

Figure 3. Poincaré domain

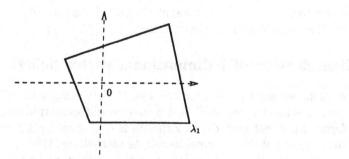

Figure 4. Siegel domain

Proof :   If the origin does not belong to the convex hull of the λ_i's, then, according to the Hahn-Banach theorem (geometrical form), there exists a line D which strictly separates $\mathcal{E}(\lambda_1, \ldots, \lambda_n)$ from 0. Then, let D' be the line passing through 0 and orthogonal to D. Let e (resp. e') a non-null vector of D (resp. D'). Even if this means changing the sign of e', we may assume that $(e', \lambda_i) > 0, 1 \leq i \leq n$ ($(.,.)$ denotes the scalar product of R^2). Let i_0 be such that $(e', \lambda_{i_0}) = \sup\{(e', \lambda_i),\ 1 \leq i \leq n\}$. Thus, there exists an integer m such that

$$\forall Q \in N^n,\ |Q| \geq m,\ \ (e', \sum_{i=1}^{n} q_i \lambda_i) > (e', \lambda_{i_0})$$

Consequently, $\forall Q \in N^n,\ |Q| \geq m$,

$$\forall 1 \leq s \leq n,\ \ \sum_{i=1}^{n} q_i \lambda_i \neq \lambda_s$$

which means that only a finite number of resonance relations can be satisfied.
□

Consequently, if the eigenvalues of the linear part of a vector field F, which is singular at the origin, belong to the Poincaré domain, then F has a polynomial normal form. In this case, we wonder what are the properties of the normalizing transformations.

Theorem 4.5 *Let F be an analytic vector field in a neighbourhood of the origin in C^n, with zero value at this point. If the eigenvalues of the Jacobian matrix of F at 0 belong to the Poincaré domain, then, F is holomorphically conjugate to its normal form.*

This theorem was proved by Poincaré, at the end of the last century, in the case where the eigenvalues were non-resonant.

5.   Classification of 1-dimensional vector fields

In this section, we shall investigate the formal classification of diffeomorphisms; this means that we shall find a class of diffeormorphisms, called normal forms, such that each diffeomorphism is conjugate, by a formal diffeomorphism, to one of these normal forms. As we shall see, this will provide the classification of differential equations in the complex plane. Following [RM83], we shall denote by F (resp. $\hat{F}$) the group of holomorphic local (resp. formal) diffeomorphisms at the origin; an element $\hat{f} \in \hat{F}$ will called **resonant** if $\hat{f}(x) = \nu x + \cdots$ where ν is a root of unity.

We shall denote by $G \subset F$ (resp. $\hat{G} \subset \hat{F}$) the sub-group of diffeomorphisms for which $\nu = 1$, that is, which are tangent to the identity.

We recall that $\hat{\mathcal{X}}$ denotes the Lie algebra of complex formal vector fields which are 1-flat at the origin, that is, which are of the form

$$\hat{X} = \hat{f}(x)\frac{d}{dx} \text{ where } \hat{f} \in C[[x]] \text{ with } \hat{f}(0) = \hat{f}'(0) = 0$$

The proposition we shall state below is concerned with vector fields with a null linear part. In fact, the linear part of the 1-dimensional vector field

$$\hat{X} = ax + \hat{f}(x)\frac{d}{dx} \text{ where } a \in C,\ \hat{f} \in C[[x]] \text{ with } \hat{f}(0) = \hat{f}'(0) = 0$$

is $x \mapsto ax$. It is resonant if and only if there exists an integer $m > 0$ such that $ma = 0$, that is $a = 0$. Consequently, if $a \neq 0$, then, according to the Poincaré-Dulac theorem, the non-linear equation can be formally linearized. Otherwise, the proposition we refer to gives a simplification of the vector field, not provided by the Poincaré-Dulac theorem, since all terms are resonant.

5.1.   Formal classification of differential equations on the plane

We recall, without proof, the following proposition, which enables us to say that conjugacy classes of $\hat{G}$ correspond to equivalence classes of $\hat{\mathcal{X}}$.

Proposition 5.1 *The exponential map* $\exp : \hat{\mathcal{X}} \to \hat{G}$ *is one-to-one.*

We shall work with vector fields and investigate the equivalent classes under the adjoint action of $\hat{g}$.

Proposition 5.2 *Each non-null vector field $\hat{X} \in \hat{\mathcal{X}}$ is equivalent, by $\hat{G}$, to one and only one of the following vector fields, called* **normal forms***:*

$$X_{\beta,k,\lambda} = \beta\frac{x^k}{1+\lambda x^k}x\frac{d}{dx}$$

where $\beta \in C^*$, k is a positive integer, and $\lambda \in C$.

Proof :   Let $\hat{X} = \hat{f}(x)\frac{d}{dx}$ be an element of $\hat{\mathcal{X}}$. It is equivalent to $X_{\beta,k,\lambda}$ by $\hat{G}$, if and only if there exists $\hat{g} \in \hat{G}$, such that

$$\hat{f}(x)\frac{d\hat{g}}{dx} = \beta\frac{\hat{g}^k}{1+\lambda\hat{g}^k}\hat{g}$$

that is

$$\left(\frac{1}{\hat{g}^{k+1}} + \frac{\lambda}{\hat{g}}\right)d\hat{g} = \beta\frac{dx}{\hat{f}}$$

After integrating and setting $\hat{g}(x) = x\hat{\phi}$ with $\hat{\phi} \in C[[x]]$ and $\hat{\phi}(0) = 1$, we set $\hat{u} = \hat{\phi}^k$; then we obtain

$$\lambda x^k \hat{u} \ln \hat{u} - 1 + kx^k \hat{u}(\lambda \ln x - \beta \int \frac{dx}{\hat{f}(x)}) = 0$$

Moreover, we can write

$$\frac{1}{\hat{f}(x)} = \frac{1}{\beta'}\left(\frac{1}{x^{k'+1}} + \frac{a_1}{x^{k'}} + \cdots + \frac{\lambda'}{x} + \cdots\right)$$

Thus, we can write

$$G(\hat{u}, x) = \lambda x^k \hat{u} \ln \hat{u} - 1 + kx^k \hat{u}((\lambda - \frac{\beta}{\beta'}\lambda') \ln x - \beta\hat{H}) = 0$$

where

$$\hat{H}(x) = \frac{1}{\beta'}\left(-\frac{1}{k'x^{k'}} - \frac{a_1}{(k'-1)x^{k'-1}} - \cdots - \frac{a_{k'-1}}{x} + a_{k+1}x + \cdots\right)$$

But we want a solution which satisfies $\hat{u}(0) = 1$; this could be possible if and only if $k = k'$ and $\beta = \beta'$, and $\lambda - \frac{\beta}{\beta'}\lambda' = 0$, that is, $\lambda = \lambda'$. The differential of G with respect to $\hat{u}$ at the point $(1, 0)$ is the identity; thus, it is invertible and, by the formal implicit function theorem, the equation $G(\hat{u}, x) = 0$ admits a unique solution $\hat{u}(x)$ such that $\hat{u}(0) = 1$. This completes the proof of the proposition.   □

Proposition 5.1 and the previous one allow us to find the conjugacy classes of $\hat{G}$:

Corollary 5.1 *Each element of $\hat{G}$, different from the identity, is conjugate, by an element of $\hat{G}$, to one and only one of the diffeomorphisms* $\exp X_{\beta,k,\lambda}$, *called* **normal forms**.

The group $\hat{F}$ is obtained, from $\hat{G}$, by adding the applications $x \mapsto ax(a \neq 0)$. Such an application transforms the vector field $X_{\beta,k,\lambda}$ into the vector field $X_{\beta',k,\lambda'}$ with $\beta' = \frac{\beta}{a^k}$ and $\lambda' = \frac{\lambda}{a^k}$. Consequently, every $\hat{X} \in \hat{\mathcal{X}}$, non-null, is equivalent, by the adjoint action of $\hat{F}$, to a unique vector field $X_{2i\pi,k,\lambda}$. The reason for the normalization $\beta = 2i\pi$ will appear in the next section.

In the same way, every diffeomorphism $\hat{g} \in \hat{G}$, different from the identity, is conjugate, by an element of $\hat{F}$, to a unique diffeomorphism $g_{2i\pi,k,\lambda} = \exp X_{2i\pi,k,\lambda}$.

In the next section, we shall set

$$X_{k,\lambda} = X_{2i\pi,k,\lambda} \text{ and } g_{k,\lambda} = g_{2i\pi,k,\lambda}$$

Let $\hat{f} \in \hat{F}$ be a formal resonant diffeomorphism, that is, $\hat{f}(x) = \nu x + \cdots$, where $\nu = e^{\frac{2i\pi p}{q}}$, p and q being relatively prime, $q \geq 2$.

Proposition 5.3 *The diffeomorphism admits the following unique decomposition $\hat{f} = \hat{\sigma} \circ \hat{n}$ where*

1. $\hat{\sigma} \in \hat{F}$ and $\hat{\sigma}^q = \underbrace{\hat{\sigma} \circ \cdots \circ \hat{\sigma}}_{q \text{ times}} = Identity$,

2. $\hat{n} \in \hat{G}$,

3. $\hat{\sigma}$ and $\hat{n}$ commute.

Proof : Let $\hat{g} = \hat{f}^q = \underbrace{\hat{f} \circ \cdots \circ \hat{f}}_{q \text{ times}}$. Hence, $\hat{g} \in G$, and by proposition 5.1, there exists $\hat{X} \in \hat{\mathcal{X}}$ such that $\hat{g} = \exp \hat{X}$. Since $\hat{f}$ and $\hat{g}$ commute, $\hat{f} * \hat{X} = \hat{X}$. Thus, $\hat{n} = \exp \frac{\hat{X}}{q}$ commutes with $\hat{f}$. If we set $\hat{\sigma} = \hat{f} \circ \hat{n}$, then we have the desired properties. Moreover, the unicity comes from the fact that $\hat{n} = \exp \frac{\hat{X}}{q}$ is the unique element of $\hat{G}$ such that $\hat{n}^q = \hat{g}$. □

Corollary 5.2 *Let $\hat{f} \in \hat{F}$ be a formal resonant diffeomorphism, $\hat{f}(x) = e^{\frac{2i p \pi}{q}} x + \cdots$. Then $\hat{f}$ is conjugate, in $\hat{F}$, to one and only one of the following elements:*

1. $\sigma(x) = e^{\frac{2i p \pi}{q}} x$ if $\hat{f}^q = Identity$,

2. $\sigma \circ g_{k,\lambda}$ where $g_{k,\lambda}$ is one of the normal forms defined previously, k being a multiple of q.

5.2. Analytic classification of resonant diffeomorphisms

In this subsection, we shall attempt to find the invariants which characterize the conjugacy classes, modulo F, of analytic resonant diffeomorphisms of the complex plane. The definition of these invariants depends on the conjugacy class, modulo $\hat{F}$, characterized by one of the normal forms obtained in the previous section. After setting some notation and stating a fundamental result, we shall study, in detail, the case of the normal form $g_{0,1}$.

We have shown that a diffeomorphism (or differential equation) of the complex plane was always formally equivalent to its normal form. We wonder whether the formal diffeomorphsim found, if it is not holomorphic in a neighbourhood of 0, could be the asymptotic expansion, in some sector of the complex plane, of an analytic function which normalizes the diffeomorphism.

5.2.1.   Asymptotic expansions and sectorial diffeomorphisms

In this section, we shall introduce some notions which are related to divergent series. We recall only the definitions needed in the next section. For a self contained and comprehensive exposition of this theory as well as the application to differential equations, we refer to the written lectures of Malgrange [Mal92] and Ramis [Ram93].

We define the **open sector** V of the complex plane, bisected by a direction d, with opening angle 2θ, and of radius $r > 0$, as the set

$$V = \{z \in C | \, |\arg z - d| < \theta, \, 0 < |z| < r\}$$

Definition 5.1 (Asymptotic expansion) *[Was65] Let V be an open sectorial domain of the complex plane (or more generally, of the Riemann surface) at 0. Let f be a holomorphic function on V and let $\hat{f} = \sum_{p=0}^{\infty} a_p x^p$ be a formal power series. $\hat{f}$ will be said to be the* **asymptotic expansion** *of f on V if, for all compact subsectors W of $V \cup \{0\}$ and for all integers n, there exists a positive constant $M_{W,n}$ such that*

$$|x|^{-n}|f(x) - \sum_{p=0}^{n-1} a_p x^p| < M_{W,n}$$

for all $x \in W$.

The set of holomorphic functions on V having an asymptotic expansion at the origin (with the operations $+$, $.$, $x^2 \frac{d}{dx}$) is a C-differential algebra denoted by $\mathcal{A}(V)$. We shall denote the Taylor jet application by $T: \mathcal{A}(V) \to C[[x]]$. We shall denote by $\mathcal{A}^{<0}(V)$ the kernel of the application T.

Definition 5.2 (Gevrey series) *[Ram93] Let $\hat{f} \in C[[z]]$ be a formal power series $\hat{f} = \sum_{n=0}^{\infty} a_n z^n$.*
$\hat{f}$ is said to be **Gevrey of order** $\frac{1}{k}$ *if*

$$\forall n \geq 0, \ |a_n| < C(n!)^{\frac{1}{k}} A^n$$

for some positive constants A and C.

Definition 5.3 (Gevrey asymptotic expansion) *[Ram93] Let V be a sectorial domain and let $f \in \mathcal{A}(V)$. Let $\hat{f} = \sum_{n=0}^{\infty} a_n z^n$ be a Poincaré asymptotic expansion of f.*
We shall say that $\hat{f}$ is a **Gevrey asymptotic expansion of order** $\frac{1}{k}$ *if for any subsector W of V, there exist positive constants A_W, C_W such that*

$$\forall z \in W, \quad |f(z) - \sum_{n=0}^{N-1} a_n z^n| < C_W (N!)^{\frac{1}{k}} A_W^N |z|^N$$

We shall denote the set of holomorphic functions on V admitting a Gevrey asymptotic expansion of order $\frac{1}{k}$ by $\mathcal{A}_{\frac{1}{k}}(V)$.

Definition 5.4 (k-summable series) *Let $k > 0$ and d be a direction. A formal power series $\hat{f}$ will be said to be k-summable in the direction d, if there exists a holomorphic function f on a sector V, bisected by d and whose opening angle is greater than $\frac{\pi}{k}$, whose Gevrey asymptotic expansion of order $\frac{1}{k}$ on V is $\hat{f}$.*
The formal power series $\hat{f}$ will be said to be k-summable, if it is k-summable in all but a finite number of directions.

Let us consider an open sectorial domain V of the complex plane C or of the Riemann surface of the logarithm. We shall also consider a covering $\{V_i\}_{i \in I}$ of V. We need to make some assumptions about the covering:

1. the covering is finite (i.e I is a finite discret set)

2. each V_i has the same radius as V

3. the intersection of any 3 elements of the covering is empty.

A covering which fulfills the above assumptions will be called a **good covering**.
We may assume that $I = [1, m]$, and we shall order the covering clockwise. We shall also write $V_{i,i+1} = V_i \cap V_{i+1}$.

Let V be an open sector at the origin. Let $G^k(V) \subset \mathcal{A}(V)$ be the set of analytic functions g on V which admit an asymptotic expansion of the form

$$\hat{g} = Tg = x + \sum_{n \geq k+1} a_n x^n$$

By the implicit function theorem, $G^k(V)$ is a group under the (non-commutative) composition law of application. We shall denote by $G^{<0}(V)$, the intersection of all $G^k(V)$, that is the set of analytic function on V which are infinitely tangent to the identity. We shall denote the set $G^1(V)$ by $G(V)$. Moreover, we shall denote by $G^{<-k}(V)$ the set of functions g such that $g - Id \in \mathcal{A}^{<-k}(V)$; this means that g is equal to the identity, up to an exponential decreasing function of order k.

We can state a theorem, due to Malgrange, and which is fundamental in the next section.

Theorem 5.3 *[Mal82][RM82] Let $\{V_i\}_{1 \le i \le n}$ be a good covering of a punctured disk by open sectors. Let $\{f_{i,j}\}$ be functions such that $f_{i,j} \in G^{<0}(V_i \cap V_j)$, if $V_i \cap V_j \ne \emptyset$. Moreover, we assume that $f_{i,j} = f_{j,i}^{-1}$.*

Then, there exist functions $f_i \in G(V_i)$, $i = 1 \ldots n$, such that

    1. the f_i's have the same asymptotic expansion,

    2. if $V_i \cap V_j \ne \emptyset$, then $f_i \circ f_j^{-1} = f_{i,j}$ on $V_i \cap V_j$.

Moreover, let $\{V_i\}$ be a good covering of a punctured disk, $\{f_i\}$ and $\{g_i\}$ be functions such that $(f_i, g_i) \in G(V_i)^2$, the f_i's admitting same asymptotic expansion $\hat{f}$ and the g_i's admitting same asymptotic expansion $\hat{g}$. Then, for all (i, j) such that $V_i \cap V_j \ne \emptyset$, there exists $h_{i,j}$ such that $h_{i,j} \circ (g_i \circ g_j^{-1}) \circ h_{i,j}^{-1} = f_i \circ f_j^{-1}$ if and only if there exists $h \in G$ such that $\hat{f} = f \circ \hat{g}$.

5.2.2. Analytic classification of diffeomorphisms formally conjugate to $g_{1,0}$

The solution $\phi(t, x)$ passing through x at $t = 0$ of the differential equation

$$\frac{du}{dt} = X_{1,0} = 2i\pi u^2$$

is

$$\phi(t, x) = \frac{x}{1 - 2i\pi t x}.$$

The "phase portrait" of the orbits of $X_{1,0}$ (corresponding to real time) is shown in figure 5.

Thus, the diffeomorphism $g = g_{1,0} = \exp X_{0,1} = \phi(1, x)$ can be written

$$g(x) = \frac{x}{1 - 2i\pi x}.$$

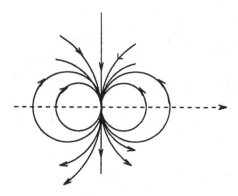

Figure 5. Orbits of the flow of $X_{1,0}$

The key point is that the function $H(x) = \exp\frac{1}{x}$ is constant along the orbit of g, that is

$$H(g^{(n)}(x)) = H(g^{(n-1)}(x)) = \cdots = H(x) = \cdots = H((g^{(-n)}(x))$$

where $g^{(n)} = \underbrace{g \circ \cdots \circ g}_{\text{n times}}$ and $g^{(-n)} = \underbrace{g^{-1} \circ \cdots \circ g^{-1}}_{\text{n times}}$.

Let $V_0 = V_{-i}(R,\theta)$ (resp. $V_1 = V_i(R,\theta)$) be the closed sector bisected by the direction $-i$ (resp. i) with radius $R > 0$ and whose opening angle is $\pi < 2\theta < 2\pi$. These two sectors offer a covering of the closed disk centred at the origin with radius R. The application $H : V_0 \setminus \{0\} \to C^*$ (resp. $H : V_1 \setminus \{0\} \to C^*$) is onto. Moreover, two points $x, x' \in V_0$ (resp. V_1) belong to the same orbit if and only if $H(x) = H(x')$. This fact means that the function H likens the quotient spaces V_0/g and V_1/g to the Riemann sphere $S^2 = C \cup \infty$, the points 0 and ∞ representing simultaneously the orbit $\{0\}$ of g.

We shall study the elements of G_d which commute with g, d being a given direction.

1. $Re(d) < 0$

   Let U be a "small" sector bisected by d, that is, all whose directions have a negative real part. Then the image of U, by H, is a neighbourhood of $0 \in S^2$. Moreover, H is a local diffeomorphism at every point but 0. Consequently, $f \in G(V)$ commutes with g if and only if there exists a local analytic diffeomorphism ϕ of S^2 at 0 such that

   $$H \circ f = \phi \circ H \qquad (16)$$

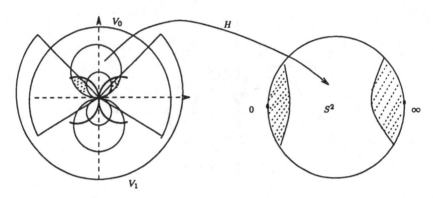

Figure 6. Isomorphisms between orbit spaces and S^2

We shall solve the preceding equation. For that purpose, we set $\phi(y) = ye^{a+\gamma(y)}$ and $f(x) = \frac{x}{1+r(x)}$, where $\gamma \in C\{y\}$, $\gamma(0) = 0$ and $r \in \mathcal{A}(U)$. Thus, the equation (16) can be written

$$\frac{1+r(x)}{x} = \frac{1}{x} + a + \gamma(e^{\frac{1}{x}})$$

that is

$$f(x) = \frac{x}{1 + ax + x\gamma(e^{\frac{1}{x}})}$$

This expression defines an element of $G(U)$; if $a = 0$, that is ϕ tangent to the identity at $0 \in S^2$, then $f \in G^{<0}(U)$. If $\gamma = 0$, then f is an element of the 1-parameter group generated by $X_{1,0}$.

2. $Re(d) > 0$

   This case is similar to the previous one, if we change neighbourhoods of 0 into neighbourhoods of $\infty \in S^2$.

3. $Re(d) = 0$

   This time, the image, under H, of any sector bisected by d is the sphere S^2. The equation (16) admits a solution $f \in G(U)$ only if ϕ is a global diffeomorphism of S^2, fixing the points 0 and ∞. Consequently, f can only be an element of the 1-parameter group of $X_{1,0}$, that is $\exp t X_{1,0}$.

The previous description can be summarized by the fact that the group $G_g^{<0}(U)$ of elements of $G^{<0}(U)$, where U is any sector whose directions have negative (resp. positive) real part, commuting with g, is isomorphic to the group of local diffeomorphisms of S^2 which are tangent to the identity, at 0 (resp. ∞). Moreover, the group $G_g^{<0}(U)$ of elements of $G^{<0}(U)$, where U is bisected by i (resp. $-i$), commuting with g, is reduced to the identity.

Let $\mathcal{U}$ be the good covering of the disk D_r consisting of the open sets $U_0 = D_r \setminus \{i\}$ and $U_1 = D_r \setminus \{-i\}$. Let us set

$$U_+ = \{x \in D_r | \ Re(x) > 0\} \text{ and } U_- = \{x \in D_r | \ Re(x) < 0\}$$

Let

$$f = (f_-, f_+) \in G_g^{<0}(U_-) \times G_g^{<0}(U_+)$$

According to the fundamental theorem 5.3, there exists $(h_0, h_1) \in G(U_0) \times G(U_1)$ such that

$$f_- = h_1^{-1} \circ h_0 \text{ on } U_- \text{ and } f_+ = h_0^{-1} \circ h_1 \text{ on } U_+$$

and $Th_0 = Th_1$, where T denotes the Taylor jet application. We shall write $\hat{f}_0$ instead of Tf_0. But both f_- and f_+ commute with g; thus,

$$h_0 \circ g \circ h_0^{-1} = h_1 \circ g \circ h_1^{-1} \text{ on } U_0 \cap U_1$$

According to the Riemann vanishing singularities theorem, $\hat{h}_0 \circ g \circ \hat{h}_0^{-1}$ is a holomorphic diffeomorphism at the origin, which is formally conjugate to g. Since h_0 and h_1 are defined up to a left composition of same analytic diffeomorphism tangent to the identity, we have associated, to each $\phi \in \text{Diff}_1(S^2, 0, \infty)$, a conjugacy class, denoted by $g(\phi)$, modulo G with formal invariants $k = 1$ and $\lambda = 0$. Moreover, the classes $g(\phi)$ and $g(\phi')$ are identical if and only if ϕ and ϕ' are conjugate by the application $y \mapsto ay$ of $S^2 (a \in C^*)$.

Eventually, we have built infinitely many analytic conjugacy classes of diffeomorphisms formally conjugate to g. Each class is characterized by a $\phi \in \text{Diff}_1(S^2, 0, \infty)$ defined up to a conjugacy by a global diffeomorphism of $(S^2, 0, \infty)$.

We shall prove that we have obtained all the analytic conjucacy classes which are formally conjugate to g. Let $g' = \hat{h} \circ g \circ \hat{h}^{-1}$, with $\hat{h} \in \hat{G}$ and $g' \in G$. According to the Borel-Ritt theorem [Was65], there exist $h_i \in \mathcal{A}(U_i)$ ($i = 0, 1$) admitting $\hat{h}$ as an asymptotic expansion at the origin in the sector of definition. We set $g'_i = h_i^{-1} \circ g' \circ h_i$. Then we have $g'_i = \psi_i \circ g$ where $\psi_i \in G^{<0}(U_i)$. Thus, according to the proposition below, there exists $f_i \in G^{<0}(U_i)$ such that $g'_i = f_i \circ g \circ f_i^{-1} (i = 0, 1)$. Eventually, we obtain

$$g' = (h_i \circ f_i) \circ g \circ (h_i \circ f_i)^{-1} \text{ on } U_i$$

The functions $h_i \circ f_i$ ($i = 0, 1$) are analytic in a sector U_i whose opening angle is greater than π and they admit $\hat{h}$ as an asymptotic expansion in U_i.

Moreover, we could show that, on the intersection $U_0 \cap U_1$, the difference $h_0 \circ f_0 - h_1 \circ f_0$ is an exponential decreasing function of order 1. Consequenly, $\hat{h}$ is 1-summable in any direction but R^+ and R^-. $\qquad \square$

The lemma we have used is known as a "sectorial normalization theorem" due to Kimura.

Proposition 5.4 (Sectorial normalization theorem) *[Mal82, RM82]*
Let $\psi \in G^{<0}(U_1)$. *Let* $\psi_p = g^{-p} \circ \psi \circ g^p$ *for* $p \geq 1$, *and* $h_p = \psi_p \circ \cdots \circ \psi_1$. *Then the sequence* (h_p) *converges to an element* $h \in G^{<0}(U_1)$ *such that,* $\psi \circ g = h^{-1} \circ g \circ h$.

Eventually, we have shown that a diffeomorphism of the complex plane, fixing the origin, could be normalized by "analytic sectorial diffeomorphisms" in sectors with sufficiently large opening angle. More generally, we have

Theorem 5.4 *[RM83, Mal82] Let* $f \in G$ *be a local analytic resonant diffeomorphism of a neighbourhood of* 0 *in the complex plane,* $f(x) = x + \cdots$. *Then* f *is conjugate, by a formal diffeomorphism* $\hat{h} \in \hat{F}$, *to a normal form* $g_{k,\lambda}$. *Then the formal power series* $\hat{h}$ *is k-summable.*

For vector fields, we have

Theorem 5.5 *Let* $X = f(x)\frac{d}{dx}$ *be a local analytic vector field in a neighbourhood of the origin of the complex plane, with* $f \in C\{x\}$, *and* $f(0) = f'(0) = 0$. *Then* X *is conjugate, by a formal diffeomorphism* $\hat{h} \in \hat{F}$, *to a normal form* $X_{k,\lambda}$. *Then the formal power series* $\hat{h}$ *is k-summable.*

6.   Classification of 2-dimensional vector fields

After having classified resonant diffeomorphisms, we may want to classify the germs, at the origin $(0,0) \in C^2$, of analytic differential equations of the form

$$\begin{cases} \frac{dx}{dz} &= P(x,y) \\ \frac{dy}{dz} &= Q(x,y) \end{cases}$$

where P and Q are both analytic in a neighbourhood of the origin $(0,0) \in C^2$. We shall assume that $P(0,0) = Q(0,0) = 0$; that is, 0 is a singular (or equilibrium) point. Thus, the above system can be rewritten in the following form:

$$\frac{dX}{dt} = AX + f(X) \tag{17}$$

where A is a complex (2×2)-matrix, $X \in C^2$, $f(0) = 0$, and $Df(0) = 0$.

6.1.   Semi-simple case

For simplicity of exposition, we shall assume that A is diagonal.

$$A = \begin{pmatrix} \lambda & 0 \\ 0 & \mu \end{pmatrix}$$

with $(\lambda, \mu) \in C^2$.

As for the diffeomorphism case, we shall be interested only in resonant cases. We distinguish two cases of resonance:

1. Degenerate case [RM82]
   This occurs when only one of the eigenvalues of A is zero.

2. Non-degenerate case [RM83]
   This occurs when λ and μ are both non-zero, and are such that

$$\frac{\lambda}{\mu} \in Q^-$$

From the geometric point of view, and by reference to the real case, in the degenerate case, we have a **saddle-node**, whereas in the non-degenerate case, we have a **resonant saddle**.

As in the case of resonant diffeomorphisms, the formal normal forms are parametrized by a finite number of parameters. The normalizing transformations, that is the changes of coordinates which transform resonant differential equations to their normal forms, are formal Gevrey series. We can show that they are k-summable in one of the two variables. For the analytic classication of saddles and foci, we refer to [MR85].

6.2.   Nilpotent case

We have shown that a 2-dimensional system with a nilpotent linear part was formally conjugate to a vector field of the form

$$\Omega_{n,p} = (y + x^p f(x))\partial_x + x^{n-1} g(x)\partial_y$$

where $(f, g) \in C[[x]]^2$, $f(0) \neq 0$ and $g(0) \neq 0$.

The integer n is a formal invariant, that is, invariant by conjugacy under a formal diffeomorphism. Considering the set Σ_1^n of vector fields having a normal form of the form $\Omega_{n,p}$ with $2p > n$, Moussu and Cerveau[MC88] have shown that a vector field of Σ_1^n was almost always conjugate to its normal form. More precisely, they proved the following theorem:

Theorem 6.1 *[MC88] Two vector fields $X \in \Sigma_1^n$ and $X' \in \Sigma_1^n$ of C^2 which are formally conjugate, are holomorphically conjugate, except in exceptional cases where: the integer n is even, the holonomy group G_X is abelian and X does not admit any holomorphic first integral.*

This work is based on the notion of *resolution of a differential form*, related to the notion of *holonomy* [MM80].

References

[AA88]   V. Arnold and D.V. Anosov. *Encyclopaedia of Mathematical Sciences: Dynamical Systems I*. Springer-Verlag, 1988.

[Arn80]   V. Arnold. *Chapitres supplémentaires de la théorie des équations différentielles ordinaires*. Mir, 1980.

[Bel78]   G.R. Belitskii. Equivalence and normal forms of germs of smooth mappings. *Russian Math. Survey*, 1978.

[Bru72]   A.D. Bruno. The analytical form of differential equations. *Trans. Mosc. Math. Soc*, 25,131-288(1971); 26,199-239(1972), 1971-1972.

[Bru89]   A.D. Bruno. *Local methods in nonlinear differential equations*. Springer Series in Soviet Mathematics, Springer-Verlag, 1989.

[CS86]   R. Cushman and J. Sanders. Nilpotent normal forms and representation theory of sl(2,R). In M. Golubitsky and J Guckenheimer, editors, *Multiparameter Bifurcation Theory*. A.M.S. Providence, 1986. Contemp. Math. 56.

[Gil]   I. Gil. Computation of the Jordan canonical form of a square matrix (using the Axiom programming langage). I.S.S.A.C.'92.

[GV]   I. Gil and G. Villard. Jordan normal form: practical experiments. To appear.

[Mal82]   B. Malgrange. Travaux d'Ecalle et de Martinet-Ramis sur les systèmes dynamiques. *Sém. Bourbaki 1981-1982*, exp. 582(1982),, 1982.

[Mal91]   B. Malgrange. *Equations différentielles à coefficients polynomiaux*. Progress in Mathematics. Birkhäuser, 1991.

[Mal92]   B. Malgrange. Séries divergentes, 1992. Cours de DEA 1991-1992, Université J. Fourier, Grenoble.

[Mar80]   J. Martinet. Normalisation des champs de vecteurs holomorphes. *Séminaire Bourbaki*, 33(1980-1981),564, 1980.

[MC88]   R. Moussu and D. Cerveau. Groupes d'automorphismes de $(C, 0)$ et équations différentielles $ydy + \cdots = 0$. *Bull. Soc. math. France*, 16(1988),459-488, 1988.

[MM80]   J.-F. Mattei and R. Moussu. Holonomie et intégrales premières. *Ann. Sci. E.N.S*, 13(1980),409-523, 1980.

[MR85]   J. Martinet and J.P. Ramis. Analytic Classification of Resonant Saddles and Foci, North-Holland, 1985.

[PdM82]   J. Palis Jr. and W. de Melo. *Geometric theory of dynamical systems: An introduction.* Springer-Verlag, 1982.

[Pos85]   M. Postinikov. *Leçons de géometrie - Groupes et algèbres de Lie.* Mir, 1985.

[Ram93]   J.P. Ramis. Divergent series and holomorphic dynamical systems. In *Bifurcations et orbites périodiques des champs de vecteurs.* Séminaires scientifiques OTAN, 1993. NATO-ASI.

[RM82]   J.P. Ramis and J. Martinet. Problèmes de modules pour des équations différentielles non linéaires du premier ordre. *I.H.E.S*, 55,63-164, 1982.

[RM83]   J.P Ramis and J. Martinet. Classification analytique des équations différentielles non linéaires résonantes du premier ordre. *Ann. Sci. E.N.S*, 4ème série,16,571-621, 1983.

[SDD91]   L. Stolovitch and J. Della Dora. Formes normales, 1991. Calcul Formel et automatique non linéaire, Lecture Notes in Control Theory, Springer-Verlag.

[Sel85]   G.R. Sell. Smooth linearization near a fixed point. *Amer. J. Math.*, 107(1985),1035-1091, 1985.

[Sib90]   Y. Sibuya. *Linear Differential Equations in the Complex Domain: Problems of Analytic Continuation.* Progress in Mathematics. Translations of Mathematical Monographs,vol. 82, A.M.S., 1990.

[Sta85]   V. Starjinski. *Méthodes appliquées en théories des oscillations non-linéaires.* Mir, 1985.

[Ste58]   S. Sternberg. On the structure of local homeomorphisms of Eu-
          clidean n-space. *Am. J. Math.*, 80,623-631(1958), 1958.

[Ste59]   S. Sternberg. The structure of local homeomorphisms. *Am. J.
          Math.*, 81,578-604(1959), 1959.

[Ste69]   S. Sternberg. *Celestial Mechanics Part II.* W.A Benjamin, 1969.

[Sto92]   L. Stolovitch. Sur les formes normales de systèmes nilpotents. *C.R.
          Acad. Sci.*, 314(1992),355-358, 1992.

[Tak73]   F. Takens. Singularities of vector fields. *I.H.E.S*, 43(1973),47-100,
          1973.

[Val]   L. Vallier. An algortithm for the computation of normal forms and
          invariant manifolds. To appear.

[Van]   A. Vanderbawhede. Center manifolds, Normal forms and Elemen-
          tary Bifurcations. Prépublication. Université de Nice.

[Var84]   V.S. Varadarajan. *Lie groups, Lie algebra, and their representa-
          tions.* Springer-Verlag, 1984.

[Was65]   W. Wasow. *Asymptotic Expansions for Ordinary Differential Equa-
          tions.* Interscience Publishers, 1965.

[Yoc85]   J.C. Yoccoz. *Centralisateurs et conjugaison différentiable des
          difféomorphismes du cercle.* Ph.D. thesis, Univerisité d'Orsay, 1985.

Versal Normal Form Computations and Representation Theory

J.A. Sanders

Contents

1. **Normal forms: a Lie algebraic formulation**   **186**

   1.1. Reductively filtered Lie algebras 186

2. **The normal form algorithms**   **190**

   2.1. Averaging methods 190

   2.2. The splitting algorithm 194

   2.3. Reduction to nilpotent normal form 198

   2.4. Nonsemisimple versal normal form 199

   2.5. Quadratic convergence 201

3. **The exponential map for vector fields**   **202**

4. **Computation of the versal normal form of an irreducible nilpotent**   **202**

5. **An example of a versal normal form computation**   **204**

1.   Normal forms: a Lie algebraic formulation

1.1.   Reductively filtered Lie algebras

In this paper we want to present an abstract normal form theory. As usual, the art of abstraction consists of retaining enough structure to formulate meaningful results and algorithms, and throw away the other structures as they only clutter one's view of things. The structure we choose to work in is something we call a reductively filtered Lie algebra, reflecting the fact that coordinate transformations lead immediately to Lie brackets on the local level.

What we want to retain, and that is where the term reductively comes in, is our ability to apply linear algebra, in particular the Jordan-Chevalley decomposition and the Jacobson-Morozov extension of the nilpotent. Apply here means that we should be able to translate the Lie algebraic problem into linear algebra, do the linear algebra and pull back the results to the Lie algebra. For instance, if we have that A is the matrix of $ad(a)$, then if $A = S + N$, there should exist elements in the Lie algebra s and n such that S is the matrix of $ad(s)$ and $a = s + n$.

Definition 1 *A* **filtered Lie algebra** $\mathcal{F}_0$ *is a Lie algebra with neighborhoods of* 0, $\mathcal{F}_k$ *such that*

   i. $\mathcal{F}_0 \supset \mathcal{F}_1 \supset \cdots \supset \mathcal{F}_k \supset \cdots$;

   ii. $\bigcap_{k=1}^{\infty} \mathcal{F}_k = 0$;

   iii. $[\mathcal{F}_k, \mathcal{F}_l] \subset \mathcal{F}_{k+l}$.

Definition 2 *The topology defined by these neighborhoods of* 0 *is called the* **filtration topology**. *To say that a process converges in the filtration topology is the same as saying that if we do a computation order by order the lower order results cannot be disturbed later by higher order ones, just as one would like it in asymptotic computations.*

Notation 1 *We let* $\mathcal{G}_k = \mathcal{F}_k / \mathcal{F}_{k+1}$.

Notation 2 *We denote for* $v \in \mathcal{F}_0$ *by* v_k *the equivalence class of* $v - \sum_{l=0}^{k-1} v_l$ *in* $\mathcal{G}_k$.

Definition 3 *A* **reductively filtered Lie algebra** *is a filtered Lie algebra with*

iv. $\dim(\mathcal{G}_k) < \infty$ *for* $k \in$;

v. $\mathcal{G}_0$ *is a reductive Lie algebra;*

vi. $\mathcal{Z}(\mathcal{G}_0)$ *consists of semisimple elements;*

vii. *Let* $v \in \mathcal{F}_0$: *if* $ad(v) : \mathcal{G}_0 \to \mathcal{G}_0$ *is semisimple (nilpotent) so is* $ad(v)$: $\mathcal{G}_k \to \mathcal{G}_k$.

Assumption 1 *In the remainder of this paper we assume that* $\mathcal{F}_0$ *is a reductively filtered Lie algebra.*

Remark 1 *The motivation for this definition is as follows:*

1. *(iv) The finite dimensionality is not always needed, but makes it possible theoretically to choose complements of subspaces. In the semisimple case one might drop this assumption, as is usually done in averaging theory.*

2. *(v) In practical computations we work in the space* $\mathcal{F}_0/\mathcal{F}_k$ *for certain* $k > 0$. *We cannot assume* $\mathcal{F}_0/\mathcal{F}_k$ *to be reductive for* $k > 1$, *since* $\mathcal{F}_1$ *is a solvable ideal, not in general in the center of* $\mathcal{F}_0/\mathcal{F}_k$. *This means that* $rad(\mathcal{F}_0/\mathcal{F}_k)$ *is larger than* $\mathcal{Z}(\mathcal{F}_0/\mathcal{F}_k)$ *and therefore* $\mathcal{F}_0/\mathcal{F}_k$ *cannot be reductive.*

3. *(v,vi) We want* $\mathcal{G}_0$ *to be reductive in order to do Jordan-Chevalley decomposition and Jacobson-Morozov extension, i.e. we want to be able to say that if* $n_0 \in \mathcal{F}_0/\mathcal{F}_1$ *is ad-nilpotent and ad-nonzero, then, since* $n_0 \notin \mathcal{Z}(\mathcal{F}_0/\mathcal{F}_1) = rad(\mathcal{F}_0/\mathcal{F}_1)$, *we have that*

$$0 \neq [n_0] \in (\mathcal{F}_0/\mathcal{F}_1)/rad(\mathcal{F}_0/\mathcal{F}_1),$$

   the last being a semisimple Lie algebra. Therefore there exist $[m_0]$ *and* $[h_0]$ *such that the triple* $\{[n_0], [m_0], [h_0]\}$ *forms a copy of* $\mathfrak{sl}_2(\ )$.

4. *(vii) Once we have done our linear algebra we want to lift the results to* $\mathcal{G}_k$. *This property will usually follow from the construction of* $\mathcal{F}_0$ *as a tensor product or function space.*

We start with a parametrized family $v(\delta) \in \mathcal{F}_0$, $\delta \in$ $^p, p \geq 0$. We want to compute the normal form of this family. To define what a normal form is, is rather complicated and we will address this issue step by step. First of all we note that $\mathcal{F}_0$ is the semidirect product of $\mathcal{G}_0$ and $\mathcal{F}_1$, i.e.

$$\mathcal{F}_0 = \mathcal{G}_0 \quad \mathcal{F}_1.$$

This allows us to work first with $\mathcal{G}_0$, and later with $\mathcal{F}_1$. Working here means: Let $t \in \mathcal{F}_0$ and compute $\exp(ad(t))v(\delta)$. It is not a priori clear that this can be computed at all. If $t \in \mathcal{F}_1$ there is no problem, since $\exp(ad(t)) = \sum_{n=0}^{\infty} \frac{1}{n!} ad^n(t)$, and $ad^n(t)v(\delta) \in \mathcal{F}_n$, so $\exp(ad(t))$ converges in the filtration topology. But if $0 \neq t \in \mathcal{G}_0$ this expression does not converge in the filtration topology. It does however converge in the usual supnorm topology for operators on finite dimensional spaces (like $\mathcal{G}_k$).

Definition 4 *We define the **weak filtration topology** as follows: a sequence $t^{(n)} \in \mathcal{F}_0$ goes to zero in the weak filtration topology if $t_k^{(n)} \overset{n \to \infty}{\to} 0$ in $\mathcal{G}_k$ (as a finite dimensional Hilbert space) for all $k \in$.*

Definition 5 *We define the **weak operator filtration topology** as follows: a sequence $t^{(n)} \in End(\mathcal{F}_0)$ goes to zero in the weak operator filtration topology if $t^{(n)}v \to 0$ in the weak filtration topology for all $v \in \mathcal{F}_0$.*

Let $t^{(n)} = \sum_{k=n}^{\infty} \frac{1}{k!} ad^k(t)$. Then this defines an operator from $\mathcal{G}_k$ into itself by the usual definition of the exponential of a matrix. It also follows that $\| t^{(n)} \| \leq \sum_{m=n}^{\infty} \frac{1}{m!} \| ad(t) \|_k^m$, where $\| \cdot \|_k$ denotes the supnorm on $End(\mathcal{G}_k)$. Thus the sequence $\{t^{(n)}\}_{n=0}^{\infty}$ converges in the weak filtration topology. This means that $\exp(ad(t)) = \sum_{n=0}^{\infty} \frac{1}{n!} ad^n(t)$ converges. It may not always be possible to carry out the explicit calculation.

Definition 6 *We say that v is **k-equivalent** to w if there exists a $t \in \mathcal{F}_0$ such that $(\exp(ad(t))v - w)_l = 0$ for $0 \leq l \leq k$.*

Definition 7 *We say that v is **equivalent** to w if there exists a $t \in \mathcal{F}_0$ such that $\exp(ad(t))v = w$.*

Definition 8 *We say that $v(\delta)$ is **versal k-equivalent** to $w(\delta)$ if there exists a family $t(\delta) \in \mathcal{F}_0$, continuous in δ, and $0 < \delta_0^{(k)} \in$ such that $(\exp(ad(t(\delta)))v(\delta) - w(\delta))_l = 0$ for $0 \leq l \leq k$ and $\| \delta \| < \delta_0^{(k)}$.*

Definition 9 *The k-equivalence class of v is called the **k-th order normal form** of v.*

Definition 10 *The equivalence class of v is called the **normal form** of v.*

Definition 11 *The versal k-equivalence class of $v(\delta)$ is called the **versal k-th order normal form** of $v(\delta)$.*

Definition 12 *If* $\delta_0 = \lim_{k \to \infty} \delta_0^{(k)}$ *exists and* $\neq 0$, *then we define the* **versal normal form** *to be the limit for* $k \to \infty$ *of the versal k-th order normal form of $v(\delta)$. (Check that this is a definition.)*

As we have seen, the easiest thing to do is to restrict the action of $\exp(ad(t))$ on $\mathcal{F}_0$ for the moment to $t \in \mathcal{F}_1$. If we do that, we can start computing immediately:

$$\exp(ad(t))v(\delta) = v(\delta) + [t, v(\delta)] + O(\mathcal{F}_2).$$

Using Notation 2 (pg. 186), we have

$$\exp(ad(t))v(\delta) = v_0(\delta) + v_1(\delta) - ad(v_0(\delta))t_1 + O(\mathcal{F}_2).$$

So the first problem will be to determine the image of $ad(v_0(\delta))|\mathcal{G}_1$. To find this image we first determine the **organizing center** of the problem. In fact, we have already fixed this, since in our definition of versal normal form the value 0 for the parameter is distinguished, since we check on $\lim_{k \to \infty} \delta_0^{(k)}$. So we take $v(0)$ to be our organizing center. In practice, one has the choice of origin in the parameter space. This choice should be determined by the following considerations:

1. $v(0)$ should be simpler than $v(\delta)$ in general; e.g. it should have its ad-eigenvalues on the imaginary axis.

2. Something interesting should happen at $\delta = 0$; e.g. a change in the topology of the linear dynamics.

Usually these two requirements do not conflict. We then treat $v(0)$ as the unperturbed problem and $v(\delta) - v(0)$ as the perturbation. Having said all this, we now go back to the Lie algebra $\mathcal{G}_0$ and find with $v(0)$ (as long as $v(0) \notin \mathcal{Z}(\mathcal{G}_0)$) a quadruple $\{s_0, n_0, m_0, h_0\}$ such that

1. $v(0) = s_0 + n_0$, $ad(s_0)$ is semisimple, $ad(n_0)$ is nilpotent and $[n_0, s_0] = 0$;

2.

$$[s_0, h_0] = [s_0, m_0] = 0,$$
$$[h_0, m_0] = 2m_0, [h_0, n_0] = -2n_0, [m_0, n_0] = h_0;$$

3. $\mathcal{G}_0 = \mathrm{im}\, ad(v(0)) \oplus (\mathrm{ker}\, ad(m_0) \cap \mathrm{ker}\, ad(s_0))$.

The definition of reductively filtered Lie algebra is such that all these properties can automatically be lifted from $\mathcal{G}_0$ to $\mathcal{G}_k$. In particular we have $\mathcal{G}_k = \mathrm{im}\, ad(v(0)) \oplus (\mathrm{ker}\, ad(m_0) \cap \mathrm{ker}\, ad(s_0))$.

2.   The normal form algorithms

2.1.   Averaging methods

To explicitly compute this splitting, we formulate first some subproblems and give algorithms to compute their solutions. Given $v(0) \in \mathcal{F}_0$, we have to solve an equation on $\mathcal{G}_k$ of the form

$$ad(v(0))t_{v(0)} = g$$

where $g \in \mathcal{G}_k$ is given and we have to determine $t_{v(0)} \in \mathcal{G}_k$ for certain $k > 0$. This problem has in general no solution, so the best we could possibly want as a result of our computations is an obstruction $\bar{g} \in \mathcal{G}_k$ and a transformation $t_{v(0)}$ such that

$$ad(v(0))t_{v(0)} = g - \bar{g}.$$

We do this by solving

$$ad(v(0))t_{v(0)}^1 = g - g^0, \qquad g^0 \in \ker ad(s_0) \tag{1}$$

first, followed by

$$ad(v(0))t_{v(0)}^0 = g^0 - g^{00}, g^{00} \in \ker ad(s_0) \bigcap \ker ad(m_0), t_{v(0)}^0 \in \ker ad(s_0)$$

and putting $\bar{g} = g^{00}, t_{v(0)} = t_{v(0)}^0 + t_{v(0)}^1$ in the end. In order to solve (1), we first turn to the problem of solving t_{s_0} in

$$ad(s_0)t_{s_0} = g - g^0, \qquad g^0 \in \ker ad(s_0). \tag{2}$$

In the following let A_k be semisimple (think of A_k as the matrix of $ad(s_0)$ or $ad(h_0)$ restricted to $\mathcal{G}_k$). We follow [LR90] , but the fact that we already obtained the Jordan decomposition simplifies the algorithm considerably. Let p be the minimal polynomial of A_k:

$$p(\lambda) = \prod_{i=1}^{k}(\lambda - \lambda_i)$$

with $\lambda_i \in$ and all different. Define

$$p_j(\lambda) = \prod_{i \neq j}(\lambda - \lambda_i)$$

and let

$$E_i = \frac{p_i(A_k)}{p'(\lambda_i)},$$

where p' is the derivative of p with respect to λ. Let f be given by a convergent power series. Then

$$f(A_k) = \sum_{i=1}^{k} f(\lambda_i)E_i.$$

Definition 13 *We define the class of quasi polynomials (in τ, say, to be denoted by $\mathfrak{P}_{\mathfrak{R}}[\tau]$) to be finite sums of the form*

$$\sum \alpha_i \tau^{\mu_i},$$

with the μ_i in some ring $\mathfrak{R}$. The ring will be often suppressed in the notation.

Using this definition this allows us to compute $\tau^{A_k} = \sum_i \tau^{\lambda_i} E_i \in \mathfrak{P}_{\mathfrak{gl}(\mathcal{G}_k)}[\tau]$, where A_k is the matrix of $ad(s_0)$. We now define $T : \mathcal{G}_k \to \mathcal{G}_k$ by

$$Tg = [\int^t \tau^{A_k} g \frac{d\tau}{\tau}]_{t=1}. \tag{3}$$

Observe that if A_k is the matrix of $ad(s_0)$ then $T = ad^{-1}(s_0)$ on $\operatorname{im} ad(s_0)$ and $T = 0$ on $\ker ad(s_0)$. This is easily checked on an eigenvector of $ad(s_0)$, and, since $ad(s_0)$ is semisimple, extends to the whole space. For let $ad(s_0)v = \mu v$. Then $\tau^{A_k} v = \tau^{\mu} v$ and

$$Tv = [\int^t \tau^{\mu} v \frac{d\tau}{\tau}]_{t=1} = [\frac{1}{\mu} t^{\mu} v]_{t=1} = \frac{1}{\mu} v$$

if $\mu \neq 0$ and if $\mu = 0$ we have

$$Tv = [\int^t v \frac{d\tau}{\tau}]_{t=1} = [\log(t) v]_{t=1} = 0.$$

Remark 2 *g^0 can be simply computed from $\tau^{A_k} g$:*

$$g^0 = Res_{\tau=0} \frac{\tau^{A_k} g}{\tau},$$

*or the constant term in $\tau^{A_k} g$. This is called the **method of averaging**.*

Lemma 2.1 *The operator T, defined by*

$$Tg = [\int^t \tau^{A_k} g \frac{d\tau}{\tau}]_{t=1},$$

obeys $ad(s_0)T = \pi_{\operatorname{im} ad(s_0)}$. If we let $t_{s_0} = Tg$ and

$$\pi^0 g = g^0 = Res_{\tau=0} \frac{\tau^{A_k} g}{\tau},$$

then we have solved the equation (2) ,

$$ad(s_0)t_{s_0} = g - g^0, \qquad g^0 \in \ker ad(s_0).$$

Proof. π^0 is the projection on $\operatorname{kerad}(s_0)$, since we take the constant term of $\tau^{A_k}g$, i.e. the term with eigenvalue zero. Thus we have

$$g = \pi_{\operatorname{imad}(s_0)}g + \pi_{\operatorname{kerad}(s_0)}g = ad(s_0)Tg + \pi^0g = ad(s_0)t_{s_0} + g^0 . \quad \Box$$

Once (2) is solved, the solution of

$$ad(s_0 + n_0)t_{v(0)}^1 = g - g^0 \tag{4}$$

is easily obtainable. For let t_{s_0} be the solution of (2) then

$$t_{v(0)}^1 = \sum_{k=0}^{\infty}(-1)^k ad^{-k}(s_0)ad^k(n_0)t_{s_0}, \tag{5}$$

which converges in the filtration topology (since $ad(n_0)$ is nilpotent on $\mathcal{G}_k$), is a solution of (4).

Lemma 2.2 *Given $g \in \mathcal{G}_k$, the equation*

$$ad(v(0))t_{v(0)}^1 = g - g^0, g^0 \in \operatorname{kerad}(s_0)$$

has a solution $(t_{v(0)}^1, g^0)$ given by

$$t_{v(0)}^1 = \sum_{k=0}^{\infty}(-1)^k ad^{-k}(s_0)ad^k(n_0)t_{s_0}$$

(where t_{s_0} is defined by Tg, see Lemma 2.1 (pg. 191)), and

$$\pi^0g = g^0 = Res_{\tau=0}\frac{\tau^{A_k}g}{\tau}.$$

Proof.

$$ad(v(0))t_{v(0)}^1 = ad(s_0 + n_0)\sum_{k=0}^{\infty}(-1)^k ad^{-k}(s_0)ad^k(n_0)t_{s_0}$$
$$= \sum_{k=0}^{\infty}(-1)^k ad^{-k}(s_0)ad^k(n_0)ad(s_0)t_{s_0}$$
$$\quad +\sum_{k=0}^{\infty}(-1)^k ad^{-k}(s_0)ad^{k+1}(n_0)t_{s_0}$$
$$= \sum_{k=0}^{\infty}(-1)^k ad^{-k}(s_0)ad^k(n_0)ad(s_0)t_{s_0}$$
$$\quad -\sum_{k=1}^{\infty}(-1)^k ad^{-k}(s_0)ad^k(n_0)ad(s_0)t_{s_0}$$
$$= ad(s_0)t_{s_0} = g - g^0 . \quad \Box$$

The last equality follows from Lemma 2.1 (pg. 191) .

Lemma 2.3 *Let $v \in \mathcal{F}_0$. Then $v = \sum_{k=0}^{\infty}v_k, v_k \in \mathcal{G}_k$. We define $\tau^{ad(s_0)}v = \sum_{k=0}^{\infty}\tau^{A_k}v_k$. If $v, w \in \mathcal{F}_0$ are $ad(s_0)$-eigenvectors with eigenvalues μ, ν, respectively, then $[v, w]$ is an eigenvector of $ad(s_0)$ with eigenvalue $\mu + \nu$.*

It follows that $\tau^{ad(s_0)}$ is a Lie algebra homomorphism, i.e.

$$\tau^{ad(s_0)}[v, w] = [\tau^{ad(s_0)}v, \tau^{ad(s_0)}w] \,, \qquad \forall v, w \in \mathcal{F}_0.$$

This gives us the possibility to carry the τ-scaling along in our computations. The advantage is in speed, since we no longer need to apply $\tau^{ad(s_0)}$ every time we want to average, the disadvantage is in memory, since we now need more space to store the extra eigenvalue information. A normal form package should include the means to switch between the two approaches, and see which works best for a given problem on a given configuration.

If we take the scaling along we have to reformulate the averaging method slightly differently, since we can no longer assume the log-term to vanish. We define two projections: the natural projection $\pi : \mathcal{G}_k \otimes \mathfrak{P}[\tau] \rightarrow \mathcal{G}$, obtained by evaluating $\tau = 1$, and $\pi^0 : \mathcal{G}_k \otimes \mathfrak{P}[\tau] \rightarrow \ker(ad(s_0)|\mathcal{G}_k)$ as follows.

$$\pi^0 g = Res_{\tau=0} \frac{g}{\tau};$$

then we define $\tilde{T} : \mathcal{G}_k \otimes \mathfrak{P} \ [\tau] \rightarrow \mathcal{G}_k \otimes \mathfrak{P} \ [\tau]$ by

$$\tilde{T}g = \int^\tau (g(\sigma) - \pi^0 g) \frac{d\sigma}{\sigma}.$$

Let v be an eigenvector of $ad(s_0)$ with eigenvalue μ, and put $\tilde{v} = \tau^\mu v$. Then there are two cases to consider:

i. $\mu \neq 0$; we have

$$\tilde{T}\tilde{v} = \int^\tau \sigma^{\mu-1} v d\sigma = \frac{1}{\mu}\tau^\mu v = \frac{1}{\mu}\tilde{v};$$

let $\pi^* ad(s_0)$ be the pull-back of $ad(s_0)$ along π; applying $\pi^* ad(s_0)$ to $\tilde{T}\tilde{v}$ gives

$$\pi^* ad(s_0)\tilde{T}\tilde{v} = \tilde{v};$$

ii. $\mu = 0$; we have, since $\tilde{v} = v = \pi\tilde{v}$,

$$\tilde{T}\tilde{v} = 0;$$

applying $\pi^* ad(s_0)$ to $\tilde{T}\tilde{v}$ gives

$$\pi^* ad(s_0)\tilde{T}\tilde{v} = 0.$$

Combining the two cases we find

$$\pi^* ad(s_0)\tilde{T} = \pi_{im\pi^* ad(s_0)}.$$

2.2.   The splitting algorithm

Using the Jacobi identity we see that $\mathcal{G}_k$ is an $\mathfrak{sl}(2;\ )$-module. We now consider the question how to explicitly compute the splitting $\mathcal{V} = \mathrm{imad}(n_0) \oplus \mathrm{kerad}(m_0)$ for a given $\mathfrak{sl}(2;\ )$-module, i.e. for given $v \in \mathcal{V}$ find $v^0 \in \mathrm{kerad}(m_0)$ and $w \in \mathrm{imad}(m_0)$ such that

$$v = v^0 + ad(n_0)t_{n_0}.$$

Remark 3 *We can take $t_{n_0} \in \mathrm{imad}(m_0)$ since for any $w \in \mathcal{V}$ we can write $w = t_{n_0} + w'$, with $w' \in \mathrm{kerad}(n_0)$ (by the dual splitting $\mathcal{V} = \mathrm{imad}(m_0) \oplus \mathrm{kerad}(n_0)$), and clearly*

$$v = v^0 + ad(n_0)w = v^0 + ad(n_0)(t_{n_0} + w') = v^0 + ad(n_0)t_{n_0}.$$

If we write $t_{n_0} = ad(m_0)v^1$ we have

$$v = v^0 + ad(n_0)ad(m_0)v^1,$$

a decomposition of v over $\mathrm{kerad}(m_0)$ and $\mathrm{imad}(m_0)$.

Let us first consider this question for an irreducible $\mathfrak{sl}(2;\ )$-module of dimension $\mu + 1$. First we want to obtain a splitting of $v \in \mathcal{V}$ along the $ad(h_0)$-eigenspaces. The spectral decomposition of $ad(h_0)$ is given by:

$$ad(h_0) = \sum \lambda_i h_i(ad(h_0)).$$

Let $v_i = h_i(ad(h_0))v$, then $v = \sum v_i$, with

$$ad(h_0)v_i = \lambda_i v_i.$$

Using the representation theory of finite dimensional $\mathfrak{sl}(2;\ )$-representations, we can now read off the position of v_i in the irreducible representation, since $\lambda_i = \mu - 2j$, or $j = \frac{1}{2}(\mu - \lambda_i)$. Let $w_j = v_i$, then

$$ad(m_0)w_j = (\mu - j + 1)w_{j-1}$$

and

$$ad(n_0)ad(m_0)w_j = (\mu - j + 1)ad(n_0)w_j = (\mu - j + 1)jw_j,$$

in other words

$$w_j = ad(n_0)w, \text{ with } w = \frac{1}{(\mu - j + 1)j}ad(m_0)w_j,$$

provided of course that $j > 0$. (One cannot have $j = \mu + 1$, since that would mean $\mu + 2$ independent vectors in a $(\mu + 1)$-dimensional representation.) If $j = 0$, then $ad(m_0)w_0 = (\mu + 1)w_{-1} = 0$ (since $w_{-1} = 0$ by construction), and $w_j = w_0 \in \mathrm{kerad}(m_0)$. We have obtained the splitting in all cases.

i. $0 < j \le \mu : v^0 = 0, w = \frac{1}{(\mu-j+1)j} w_j$;

ii. $j = 0 : v^0 = w_j, w = 0$.

All this assumes that we knew μ beforehand. Now suppose we know that the representation is irreducible, but we do not know its dimension. Then we can read off the dimension from the element in the representation in $\mathrm{kerad}(m_0)$. For if $w_0 \in \mathrm{kerad}(m_0)$, then we have $ad(h_0)w_0 = \mu w_0$, and the dimension is $\mu + 1$. So, how do we get an element in $\mathrm{kerad}(m_0)$? Well, by applying the nilpotent operator $ad(m_0)$ to $v \in \mathcal{V}$ until the result is zero. Let $w^{(0)} = v$ and $w^{(j+1)} = ad(m_0)w^{(j)}$. Let k be such that $w^{(k)} \ne 0$, but $ad(m_0)w^{(k)} = 0$. Then compute μ from $ad(h_0)w^{(k)} = \mu w^{(k)}$. We are now in the situation we have treated first.

We now give an alternative procedure which does not require the immediate application of the spectral decomposition and which also works in the reducible case.

Take $v \in \mathcal{V}$. Let $v^{(0)} = v$ and $v^{(j+1)} = ad(m_0)v^{(j)}$. As before, let k be such that $0 \ne v^{(k)} \in \mathrm{kerad}(m_0)$. Now apply the spectral decomposition on $v^{(k)}$:

$$v^{(k)} = \sum v_{i,0}^{(k)},$$

with $ad(h_0)v_{i,0}^{(k)} = \lambda_i v_{i,0}^{(k)}$. Let

$$w^{(k-1)} = \sum\nolimits_{\lambda_i \ne 0} \lambda_i^{-1} ad(n_0)v_{i,0}^{(k)}.$$

Then

$$ad(m_0)w^{(k-1)} = \sum\nolimits_{\lambda_i \ne 0} \lambda_i^{-1} ad(m_0)ad(n_0)v_{i,0}^{(k)}$$
$$= \sum\nolimits_{\lambda_i \ne 0} \lambda_i^{-1} (ad(h_0) + ad(n_0)ad(m_0))v_{i,0}^{(k)}$$
$$= \sum\nolimits_{\lambda_i \ne 0} \lambda_i^{-1} ad(h_0)v_{i,0}^{(k)}$$
$$= \sum\nolimits_{\lambda_i \ne 0} v_{i,0}^{(k)}$$

and this implies, since $v^{(k)} = ad(m_0)v^{(k-1)}$, that

$$v^{(k-1)} = w^{(k-1)} \bmod \mathrm{kerad}(m_0).$$

In other words,

$$v^{(k-1)} = ad(n_0)\sum\nolimits_{\lambda_i \ne 0} \lambda_i^{-1} v_{i,0}^{(k)} \bmod \mathrm{kerad}(m_0).$$

and this gives the required splitting on level $k - 1$. In general, if we have

$$v^{(k-j)} = \sum v_{i,j}^{(k)}$$

with

$$ad(h_0)v_{i,j}^{(k)} = (\lambda_i - 2j)v_{i,j}^{(k)},$$
$$ad(n_0)v_{i,j}^{(k)} = (j+1)v_{i,j+1}^{(k)},$$
$$ad(m_0)v_{i,j}^{(k)} = (\lambda_i - j + 1)v_{i,j-1}^{(k)},$$

let $w^{(k-j-1)} = \sum \frac{1}{(j+1)(\lambda_i-j)} ad(n_0)v_{i,j}^{(k)}$, then

$$ad(m_0)w^{(k-j-1)} = \sum \frac{1}{(j+1)(\lambda_i-j)} ad(m_0)ad(n_0)v_{i,j}^{(k)}$$
$$= \sum \frac{1}{(j+1)(\lambda_i-j)} (ad(h_0) + ad(n_0)ad(m_0))v_{i,j}^{(k)}$$
$$= \sum \frac{1}{(j+1)(\lambda_i-j)} (\lambda_i - 2j + (\lambda_i - j + 1)j)v_{i,j}^{(k)}$$
$$= \sum v_{i,j}^{(k)}.$$

Observe that we could have obtained the factor $\frac{1}{(j+1)(\lambda_i-j)}$ by scaling $v_{i,j}^{(k)}$ by $X^j Y^{\lambda_i-j}$ and applying the following operation:

$$\mathcal{I} : X^j Y^{\lambda_i-j} \mapsto \frac{1}{Y}\int^Y \frac{1}{\eta}\int^X \xi^j \eta^{\lambda_i-j} d\xi d\eta$$
$$= \frac{1}{Y}\int^Y \frac{1}{j+1} X^{j+1} \eta^{\lambda_i-j-1} d\eta$$
$$= \frac{1}{(j+1)(\lambda_i-j)} X^{j+1} Y^{\lambda_i-j-1}.$$

The resulting scaling with $X^{j+1}Y^{\lambda_i-j-1}$ reflects the position of the object $ad(n_0)v_{i,j}^{(k)}$ in the representation. Observe that the total power (of X and Y) remains λ_i. This indicates how we can codify the information on the position of each eigenvector in every irreducible component by polynomials in a uniform way. The simplest example of this is the application of $Y^{ad(h_0)}$ on $\ker ad(m_0)$: if $v^{(k)} \in \ker ad(m_0)$ then $Y^{ad(h_0)}v$ is scaled and ready to be used. Applying both the double integration and $ad(n_0)$ to $Y^{ad(h_0)}v$ we obtain $v^{(k-1)}$, but now scaled, modulo a term in $\ker ad(m_0)$, which can be scaled by applying $Y^{ad(h_0)}$ to it. Thus we have lifted the scaling one level and this is the basis of a recursive algorithm which we describe more formally now.

We denote for any given $v \in \mathcal{V}$ the required codification with X and Y by $\mathcal{C}v$, so $\mathcal{C} : \mathcal{V} \to \mathcal{V} \otimes P[X,Y]$. We have the following equation for $\mathcal{C}$:

$$\mathcal{C}v = ad(n_0)\mathcal{I}\mathcal{C}ad(m_0)v + Y^{ad(h_0)}(v - ad(n_0)\mathcal{I}\mathcal{C}ad(m_0)v|_{(X=1,Y=1)}), (6)$$

where we remark that if we let

$$t_{n_0} = \mathcal{I}\mathcal{C}ad(m_0)v|_{(X=1,Y=1)} \qquad (7)$$

and

$$v^0 = ad(n_0)\mathcal{I}Cad(m_0)v|_{(X=1,Y=1)} \in \ker ad(m_0), \qquad (8)$$

then

$$v = v^0 + ad(n_0)t_{n_0}.$$

(6) allows us to compute Cv recursively. The recursion ends because $ad(m_0)$ is nilpotent on $\mathcal{V}$, and the result must be right because the equation holds for the $ad(h_0)$-eigenvectors, which span $\mathcal{V}$. We have then solved the purely nilpotent splitting problem.

Lemma 2.4 *For given $v \in \mathcal{G}_k, k > 0$, there exists a solution $t_{n_0} \in \mathcal{G}_k$ to the problem*

$$ad(n_0)t_{n_0} = v - v^0$$

where $v^0 \in \ker ad(m_0) \cap \mathcal{G}_k$. Furthermore v^0 can be computed from v by (8) and t_{n_0} by (7).

Proof. Take $v_{i,j}^{(k)}$. If $j = 0$ the right hand side of (6) evaluates to $Y^{ad(h_0)}v_{i,j}^{(k)} = Y^{\lambda_i}v_{i,j}^{(k)} = Cv_{i,j}^{(k)}$. If $j > 0$ we have

$$ad(n_0)\mathcal{I}Cad(m_0)v_{i,j}^{(k)} =$$
$$= (\lambda_i - j + 1)ad(n_0)\mathcal{I}Cv_{i,j-1}^{(k)}$$
$$= (\lambda_i - j + 1)ad(n_0)\mathcal{I}X^{j-1}Y^{\lambda_i-j+1}v_{i,j-1}^{(k)}$$
$$= (\lambda_i - j + 1)ad(n_0)\frac{1}{Y}\int^Y \frac{1}{\eta}\int^X \xi^{j-1}\eta^{\lambda_i-j+1}d\xi d\eta v_{i,j-1}^{(k)}$$
$$= (\lambda_i - j + 1)ad(n_0)\frac{1}{Y}\int^Y \frac{1}{j}X^j\eta^{\lambda_i-j}d\eta v_{i,j-1}^{(k)}$$
$$= ad(n_0)\frac{1}{j}X^jY^{\lambda_i-j}v_{i,j-1}^{(k)}$$
$$= X^jY^{\lambda_i-j}v_{i,j}^{(k)} = Cv_{i,j}^{(k)}.$$

Restricting to $\ker ad(s_0)$ and taking g^0 for v, we have obtained

$$g^0 = g^{00} + ad(n_0)t^0_{v(0)},$$

with $g^{00} \in \ker ad(n_0) \cap \ker ad(m_0)$ and $t^0_{v(0)} \in \ker ad(n_0)$. Or

$$g^0 = g^{00} + ad(v(0))t^0_{v(0)}.$$

This concludes our discussion, since now we have

$$g = g^{00} + ad(v(0))(t^0_{v(0)} + t^1_{v(0)}),$$

the desired splitting.

2.3.   Reduction to nilpotent normal form

First we consider the problem

$$ad(v(\delta))t^1_{v(\delta)} = g - g^0, \quad g^0 \in \ker ad(m_0).$$

We know from Lemma 2.4 (pg. 197) how to solve for given $g \in \mathcal{G}_k$ the equation

$$ad(n_0)t^1_{n_0} = g - g^0,$$

and we denote this $t^1_{n_0}$ by $\overline{N}g$. We have

$$ad(n_0) \circ \overline{N} = \pi_{\operatorname{im} ad(n_0)} = 1 - \pi_{\ker ad(m_0)}.$$

If g has $ad(h_0)$-eigenvalue λ, then $\overline{N}g$ has $ad(h_0)$-eigenvalue $\lambda + 2$, i.e. the $ad(h_0)$-degree of $\overline{N}$ is 2. Let $V(\delta) = v(\delta) - v(0)$.

Assume $V(\delta) \in \ker ad(m_0) \cap \ker ad(s_0)$; therefore it commutes with s_0 and m_0, and can be written as a sum of $ad(h_0)$-eigenvectors with nonnegative eigenvalues, i.e. the $ad(h_0)$-degree of all terms in $V(\delta)$ is ≥ 0. Since s_0 commutes with both h_0 and m_0, its $ad(h_0)$-eigenvalue is zero, i.e. the $ad(h_0)$-degree of $ad(s_0)$ is 0. This implies that the $ad(h_0)$-degree of $ad(s_0+V(\delta))\circ \overline{N}$ is ≥ 2, implying that this operator is nilpotent on $\mathcal{G}_k$. We denote this nilpotent operator by $Q = ad(s_0 + V(\delta)) \circ \overline{N}$. This allows us to define

$$t^1_{n_0} = \overline{N} \circ (1+Q)^{-1}g. \tag{9}$$

Then

$$ad(s_0 + n_0 + V(\delta))t^1_{n_0}$$
$$= ad(s_0 + n_0 + V(\delta)) \circ \overline{N} \circ (1+Q)^{-1}g$$
$$= (ad(n_0) \circ \overline{N} + Q) \circ (1+Q)^{-1}g$$
$$= (1 + Q - (1 - ad(n_0) \circ \overline{N}))(1+Q)^{-1}g$$
$$= g - (1 - ad(n_0) \circ \overline{N})(1+Q)^{-1}g$$
$$= g - \pi_{\ker ad(m_0)}(1+Q)^{-1}g.$$

We define

$$g^0 = (1 - ad(n_0) \circ \overline{N})(1+Q)^{-1}g \tag{10}$$
$$= \pi_{\ker ad(m_0)}(1+Q)^{-1}g.$$

Lemma 2.5 *For given $g \in \mathcal{G}_k, k > 0$, there exists a solution $t_{n_0} \in \mathcal{G}_k$ to the problem*

$$ad(v(\delta))t^1_{n_0} = g - g^0$$

where $g^0 \in \ker ad(m_0) \cap \mathcal{G}_k$. Furthermore g^0 can be computed from g by (10) and $t^1_{n_0}$ by (9).

2.4.   Nonsemisimple versal normal form

We consider the problem

$$ad(v(\delta))t^0_{v(\delta)} = g^0 - g^{00}$$

where $g^0 \in \ker \overline{N} = \ker ad(m_0)$ and $g^{00} \in \ker ad(m_0) \cap \ker ad(s_0)$. We take g^{00} to be the projection of g^0 on $\ker ad(s_0)$ (the average of g^0). This makes $ad(s_0)$ invertible on $g^0 - g^{00}$. Then the inverse of $ad(s_0 + n_0)$ exists too. We define operators $C(\delta)$ and $K(\delta)$ such that $K(\delta) = ad(v(\delta))C(\delta) : \ker ad(m_0) \rightarrow \ker ad(m_0)$ and $K(\delta) \neq id_{\ker \overline{N}}$ as long as $\delta \neq 0$ for $g \in im\, ad(s_0)$ as follows: Let $P = \overline{N} \circ ad(s_0 + V(\delta))$. And put

$$C(\delta)g = (1 + P)^{-1}(1 - \overline{N} \circ ad(n_0))ad^{-1}(s_0 + n_0)g$$

Check that $C(\delta)$ is well defined. We now prove that $K(\delta)$ maps into $\ker \overline{N}$. Indeed, let $g \in \ker \overline{N}$. Then

$$
\begin{aligned}
& ad(v(\delta)) \circ C(\delta)g \\
&= ad(v(\delta)) \circ (1 + P)^{-1} \circ (1 - \overline{N} \circ ad(n_0)) \circ ad^{-1}(s_0 + n_0)g \\
&= ad(v(\delta)) \circ (1 + P)^{-1} \circ (1 + P - \overline{N} \circ ad(v(\delta))) \circ ad^{-1}(s_0 + n_0)g \\
&= ad(v(\delta)) \circ (1 - (1 + P)^{-1} \circ \overline{N} \circ ad(v(\delta))) \circ ad^{-1}(s_0 + n_0)g \\
&= (1 - ad(v(\delta))(1 + P)^{-1} \circ \overline{N}) \circ ad(v(\delta)) \circ ad^{-1}(s_0 + n_0)g \\
&= (1 - ad(v(\delta)) \circ \overline{N} \circ (1 + Q)^{-1}) \circ ad(v(\delta)) \circ ad^{-1}(s_0 + n_0)g \\
&= (1 - (1 - \pi_{\ker \overline{N}} \circ (1 + Q)^{-1})) \circ ad(v(\delta)) \circ ad^{-1}(s_0 + n_0)g \\
&= \pi_{\ker \overline{N}} \circ (1 + Q)^{-1} \circ ad(v(\delta)) \circ ad^{-1}(s_0 + n_0)g.
\end{aligned}
$$

Now let

$$A_m =< g^0, K(\delta)g^0, \cdots, K(\delta)^m g^0 >$$

until for some $m \geq 0$,

$$K(\delta)^{m+1} g^0 \in A_m.$$

We write

$$K(\delta)^{m+1} g^0 = \sum_{j=0}^{m} a_j K(\delta)^j g^0$$

and form the matrix $B(\delta) = K(\delta)|A_m$. Then we have

$$
\begin{pmatrix} g^0 \\ \cdot \\ \cdot \\ \cdot \\ K(\delta)^m g^0 \end{pmatrix} = ad(v(\delta)) \circ B(\delta)^{-1} \begin{pmatrix} C(\delta)g^0 \\ \cdot \\ \cdot \\ \cdot \\ C(\delta)K(\delta)^m g^0 \end{pmatrix}.
$$

Thus we can compute w such that

$$ad(v(\delta))t^0_{v(\delta)} = g^0.$$

We have to check that in the process we did not divide by δ. This could have happened if $\det(B(0)) = 0$. We can show that

$$K(\delta)|\ker\overline{N}$$
$$= ad(v(\delta)) \circ (1 + P)^{-1} \circ (1 + \overline{N} \circ ad(s_0)) \circ ad^{-1}(s_0 + n_0),$$

and it follows that $B(0) = id_{A_m}$, hence has determinant 1.

Lemma 2.6 *For given $g^0 \in \ker ad(m_0) \cap \mathcal{G}_k, k > 0$, there exists a constant $\delta_0^{(k)} > 0$ and a solution $t^0_{v(\delta)} \in \mathcal{F}_l(\ ^n)$ to the problem*

$$ad(v(\delta))t^0_{v(\delta)} = g^0 - g^{00}$$

where $g^{00} \in \ker ad(m_0) \cap \ker ad(s_0) \cap \mathcal{F}_l(\ ^n)$ for $|\delta| < \delta_0^{(k)}$. Furthermore g^{00} can be computed from g^0 by averaging and $t^0_{v(\delta)}$ is the first component of

$$B(\delta)^{-1}C(\delta)\begin{pmatrix} g^0 - g^{00} \\ \cdot \\ \cdot \\ \cdot \\ K(\delta)^m(g^0 - g^{00}) \end{pmatrix}.$$

Remark 4 *The $\delta_0^{(k)}$ arises because we have to invert $B(\delta)$; this may not be possible for certain discrete values of δ, which we call **resonances**. Clearly there exists the possibility that*

$$\lim_{l\to\infty}\delta_0^{(k)} = 0.$$

Remark 5 *We should mention here that this is the first time that the method is not as direct as one might wish, since we have to invert a matrix, something we try to avoid doing symbolically in high dimensional spaces. The given method keeps the dimension minimal. In a typical example (from the L_4-problem, Cf [CK86]) it reduces the dimension from 24 to 2.*

Theorem 2.7 *For given $g \in \mathcal{G}_k, k > 0$, there exists a constant $\delta_0^{(k)} > 0$ and a solution $t_{v(\delta)} \in \mathcal{G}_k$ to the problem*

$$ad(v(\delta))t_{v(\delta)} = g - g^{00}$$

with $g^{00} \in \ker ad(m_0) \cap \ker ad(s_0) \cap \mathcal{G}_k$ for $|\delta| < \delta_0^{(k)}$.

Proof. Combine Lemma 2.5 (pg. 198) and Lemma 2.6 (pg. 200) , writing $g - g^{00} = (g - g^0) + (g^0 - g^{00})$, and take $t_{v(\delta)} = t^0_{v(\delta)} + t^1_{v(\delta)}$, using linearity.

2.5.   Quadratic convergence

Remark 6 *Nowhere have we used the fact that $V(\delta)$ is linear. It might as well have been a nonlinear vector field in normal form.*

Instead of taking $\mathcal{G}_l$ we take a finite direct sum of such spaces:

$$\oplus_{l=2^k}^{2^{k+1}-1}\mathcal{G}_l,$$

on which space $ad_{s_0+n_0+V(\delta)}$ acts if we forget the terms of order $\geq 2^{k+1}$.

Suppose we have the following situation:

$$v = v(0) + \sum_{l=0}^{2^k-1}\overline{v}_l + \sum_{l=2^k}^{\infty}v_l, \quad k \in \quad , \tag{11}$$

where the $\overline{v}_l$ are in normal form, i.e. elements of $\ker ad_{m_0} \cap \ker ad_{s_0} \cap \mathcal{G}_l$. (Strictly speaking this last assumption is only necessary for $l = 0$. But it would not make much sense to normalize higher order terms when the lower order terms are not in normal form.) For $k = 0$ this corresponds to the situation that the linear part is in versal normal form, and the nonlinear terms have yet to be normalized. We associate with $v(0)$ the matrix $s_0 + n_0$ and with $\overline{v}_0$ the matrix $V(\delta)$. We can now solve the problem

$$ad_{v(0)+\sum_{l=0}^{2^k-1}\overline{v}_l}w = \sum_{l=2^k}^{2^{k+1}-1}(v_l - \overline{v}_l) \quad mod\mathcal{G}_{2^{k+1}},$$
$$w \in \oplus_{l=2^k}^{2^{k+1}-1}\mathcal{G}_l$$

using Theorem 2.7 (pg. 200) and Remark 6 (pg. 201) . Then

$$\exp(ad(w))v = v + ad(w)v \quad mod\mathcal{F}_{2^{k+1}}$$
$$= v(0) + \sum_{l=0}^{2^k-1}\overline{v}_l + \sum_{l=2^k}^{2^{k+1}-1}v_l - ad_{v(0)+\sum_{l=0}^{2^k-1}\overline{v}_l}w \quad mod\mathcal{F}_{2^{k+1}}$$
$$= v(0) + \sum_{l=0}^{2^k-1}\overline{v}_l + \sum_{l=2^k}^{2^{k+1}-1}v_l - \sum_{l=2^k}^{2^{k+1}-1}(v_l - \overline{v}_l) \quad mod\mathcal{F}_{2^{k+1}}$$
$$= v(0) + \sum_{l=0}^{2^{k+1}-1}\overline{v}_l \quad mod\mathcal{F}_{2^{k+1}}.$$

This gives us quadratic convergence of the normalization process in the filtration topology since the order of the terms that are possibly not in normal form is twice as high as it was before in (11). We sum this up in

Theorem 2.8 *A δ-family of elements in a reductively filtered Lie algebra with its zeroth order part in versal normal form can be brought into higher order normal form with a computation that converges quadratically, with the possible exception of a finite number of nonzero δ-values.*

3.   The exponential map for vector fields

In this section we compute

$$\exp(tad(x_i\tfrac{\partial}{\partial x_j}))\sum_{k,l}a_{k,l}x_k\tfrac{\partial}{\partial x_l}, \qquad x\in{}^n.$$

Let

$$E_{ij}=x_i\tfrac{\partial}{\partial x_j}.$$

Then

$$ad(E_{ij})E_{kl}=[E_{ij},E_{kl}]=\delta_{jk}E_{il}-\delta_{il}E_{kj},$$
$$ad^2(E_{ij})E_{kl}=[E_{ij},\delta_{jk}E_{il}-\delta_{il}E_{kj}]=\delta_{jk}\delta_{ji}E_{il}+\delta_{il}\delta_{ij}E_{kj}-2\delta_{jk}\delta_{il}E_{ij},$$
$$ad^3(E_{ij})E_{kl}=[E_{ij},\delta_{jk}\delta_{ji}E_{il}+\delta_{il}\delta_{ij}E_{kj}-2\delta_{jk}\delta_{il}E_{ij}]$$
$$=\delta_{ij}(\delta_{jk}E_{il}-\delta_{il}E_{kj})=\delta_{ij}ad(E_{ij})E_{kl}.$$

This implies for $i\neq j$ that

$$\exp(tad(E_{ij}))E_{kl}=E_{kl}+t(\delta_{jk}E_{il}-\delta_{il}E_{kj})-t^2\delta_{jk}\delta_{il}E_{ij}. \qquad (12)$$

For the diagonal elements we compute the action of $\sum_i t_i E_{ii}$. We find

$$ad(\sum_i t_i E_{ii})E_{kl}=\sum_i t_i[E_{ii},E_{kl}]$$
$$=\sum_i t_i(\delta_{ik}E_{il}-\delta_{il}E_{ki})=(t_k-t_l)E_{kl}$$

and

$$\exp(ad(\sum_i t_i E_{ii}))\sum_{kl}a_{kl}E_{kl}=\sum_{kl}a_{kl}e^{t_k-t_l}E_{kl}. \qquad (13)$$

For $i\neq j$ we have

$$\exp(tad(E_{ij}))v=v+t\sum_k(a_{jk}E_{ik}-a_{ki}E_{kj})-t^2a_{ji}E_{ij}. \qquad (14)$$

This enables us to do the linear algebra necessary to put the linear vector field into versal normal form.

4.   Computation of the versal normal form of an irreducible nilpotent

Remark 7 *The following is a coordinate dependent method, in clear contrast to the methods we have been developing thus far. As such, it should not be considered as a final result, but as an intermediate one, to be improved later.*

We suppose that the coefficients are elements in the local ring $(\mathfrak{R}, \mathfrak{M}, \ )$, i.e. $\mathfrak{M}$ the unique maximal ideal of noninvertible elements, and the residue field $\mathfrak{R}/\mathfrak{M} = \ $. We restrict out attention to the situation with $\mathcal{G}_0 = \mathfrak{gl}(n, \mathfrak{R})$ (or possibly $\mathcal{G}_0 = \mathfrak{sl}(n, \mathfrak{R})$). We assume now that we have the equivalent of an irreducible nilpotent block, i.e.

$$a_{i+1,i} \notin \mathfrak{M}, i = 1, \cdots, n-1,$$
$$a_{i,j} \in \mathfrak{M}, i \neq j+1, i, j = 1, \cdots, n.$$

We use induction (using lexicographic ordering) to prove that we can put $v = \sum_{kl} a_{kl} E_{kl}$ in versal normal form. We say that $(k, l) \prec (i, j)$ if $l < j$ or if $l = j$ and $k > i$. Fix q, p with $p + 1 < q$ and assume that $a_{i,j} = 0$ for $(i, j) \prec (q, p)$ or, more explicitly, for $j < p$ and $j + 1 < i \leq n$ and for $j = p$ and $q < i \leq n$. For the lowest element (with respect to $\prec$) in our index set (i.e. $q = n, p = 1$) this condition is empty. We want to remove the E_{qp} term, i.e. make $a_{q,p}$ equal to zero, without disturbing our initial assumption. Using (14) we find

$$\exp(t\,\text{ad}(E_{q,p+1}))v = v + t\sum_k (a_{p+1,k} E_{q,k} - a_{k,q} E_{k,p+1}) - t^2 a_{p+1,q} E_{q,p+1}.$$

The coefficient of $E_{q,p}$ is

$$a_{q,p} + t a_{p+1,p}$$

and since $a_{p+1,p}$ is invertible by assumption, we can take $t = -a_{p+1,p}^{-1} a_{q,p}$ to make the coefficient equal to zero. We have to check that we did not disturb the induction hypothesis. The only possible disturbance is the term

$$\sum_k^{p-1} a_{p+1,k} E_{q,k}.$$

But here we have $k < p$ and $k + 1 < p + 1 \leq n$, and by hypothesis $a_{p+1,k} = 0$ for this range. We have now shown that we can remove all terms of the form $E_{q,p}$ with $p + 1 < q$.

We now turn to the terms $E_{p+1,p}$ for $p = 1, \cdots, n - 1$. Using (13) we find

$$\exp(\text{ad}(\sum_i t_i E_{ii}))\sum_{kl} a_{kl} E_{kl} = \sum_{kl} a_{kl} e^{t_k - t_l} E_{kl}.$$

The coefficients of $E_{p+1,p}$ are

$$e^{t_{p+1} - t_p} a_{p+1,p}$$

and to make them equal to a fixed $r_p \in \ $ we put $t_n = 0$ and

$$t_p = t_{p+1} + \log(r_p^{-1} a_{p+1,p}).$$

(Alternatively, we might require the sum of the t_i to vanish, thus restricting our transformation elements to $\mathfrak{sl}(n, \ )$. This would imply that we take

$$t_n = \frac{1}{n-1}\sum_{p=1}^{n-1}p\log(r_p^{-1}a_{p+1,p})$$

and

$$t_p = t_{p+1} + \log(r_p^{-1}a_{p+1,p}).)$$

This works as long as $r_p^{-1}a_{p+1,p} > 0$. Clearly this does not allow us to reach the Jordan normal form, since if coefficients are supposed to be real and negative, one could never make them equal to 1. From the point of view of normal form theory this seems the most natural approach, since the Jordan normal form is not necessary for our theory. The existence of the log function might put some restrictions on the kind of coefficient ring that is allowed, but this is at most of theoretical interest, and no attempt will be made to make this more substantial. One interpretation would be to consider convergence in the $\mathfrak{m}$-adic topology, but we might be able to obtain nonformal results.

The remaining terms are upper triangular and we can use the splitting algorithm to further normalize v. The grading induced by the nilpotent block is:

$$gr(E_{i,j}) = 2(j-i).$$

Since the transformation has a grading which is the grading of the object to be removed plus two, its grading is strictly positive, ensuring the convergence in the filtration topology of $\exp(ad(transformation))$. This concludes our analysis of the irreducible nilpotent block. We have shown how to compute its versal deformation.

5.   An example of a versal normal form computation

Let us take $n = 3$ and

$$v = E_{21} + E_{32} + \delta E_{31} + \alpha E_{33}.$$

The local ring $\mathfrak{R}$ consists of the intersection of the rational functions and formal power series in α and δ. Thus $\frac{1}{1+\delta\alpha} \in \mathfrak{R}$.

Then the first transformation (to remove the δE_{31} term) is $-\delta E_{32}$, and we find

$$v^{(1)} = \exp(ad(-\delta E_{32}))v = E_{21} + E_{32} + \delta\alpha E_{32} + \alpha E_{33}.$$

Next we want to make the components of E_{21} and E_{32} equal to 1 and 2, respectively. We can do this by choosing

$$t_1 = t_2 = 0, t_3 = -\log((1 + \delta\alpha)/2),$$

to obtain

$$v^{(2)} = \exp(ad(-\log((1 + \delta\alpha)/2)E_{33}))v^{(1)} = E_{21} + 2E_{32} + \alpha E_{33}.$$

Denoting $E_{21} + 2E_{32}$ by n_0, and forming $m_0 = 2E_{12} + E_{23}$, we can define $h_0 = [m_0, n_0] = 2(E_{11} - E_{33})$. We have now a copy of $\mathfrak{sl}(2,) = < n_0, m_0, h_0 >$, with

$$[h_0, m_0] = 2m_0,$$
$$[h_0, n_0] = -2n_0,$$

and

$$[m_0, n_0] = h_0.$$

Applying $\tau^{ad(h_0)}$ to $E_{k,l}$ gives us $\tau^{t_k - t_l} E_{kl}$, with $t_1 = 2, t_2 = 0$ and $t_3 = -2$. So $\tau^{ad(h_0)} E_{13} = \tau^4 E_{13}$. For the splitting algorithm one has to compute the images of the object under consideration, in our case αE_{33}, under the application of $ad(m_0)$. Let $w^{(0)} = E_{33}$. Then we compute $w^{(1)} = ad(m_0)w^{(0)} = [2E_{12} + E_{23}, E_{33}] = E_{23}$, and $w^{(2)} = ad(m_0)w^{(1)} = [2E_{12} + E_{23}, E_{23}] = 2E_{13}$. Finally $w^{(3)} = ad(m_0)w^{(2)} = [2E_{12} + E_{23}, 2E_{13}] = 0$. Then

$$\mathcal{C}w^{(2)} = Y^{ad(h_0)}w^{(2)} = 2Y^4 E_{13}.$$

This allows us to compute

$$ad(n_0)\mathcal{I}\mathcal{C}ad(m_0)w^{(1)} = ad(n_0)\mathcal{I}\mathcal{C}w^{(2)} = 2ad(n_0)\mathcal{I}Y^4 E_{13}$$
$$= \frac{2}{Y}\int^Y \frac{1}{\eta}\int^X \eta^4 d\xi d\eta [E_{21} + 2E_{32}, E_{13}]$$
$$= \frac{2}{Y}\int^Y X\eta^3 d\eta (E_{23} - 2E_{12})$$
$$= \frac{1}{2}XY^3(E_{23} - 2E_{12}).$$

This implies

$$\mathcal{C}w^{(1)} = \frac{1}{2}XY^3(E_{23} - 2E_{12}) + Y^{ad(h_0)}(E_{23} - \frac{1}{2}(E_{23} - 2E_{12}))$$
$$= \frac{1}{2}XY^3(E_{23} - 2E_{12}) + Y^2(\frac{1}{2}E_{23} + E_{12})$$
$$= \frac{1}{2}(XY^3 + Y^2)E_{23} + (-XY^3 + Y^2)E_{12}.$$

This allows us to do the last step:

$$
\begin{aligned}
ad(n_0)&\mathcal{I}Cad(m_0)w^{(0)} = ad(n_0)\mathcal{I}Cw^{(1)} \\
&= ad(n_0)\mathcal{I}(\tfrac{1}{2}(XY^3 + Y^2)E_{23} + (-XY^3 + Y^2)E_{12}) \\
&= \frac{1}{Y}\int^{Y}\frac{1}{\eta}\int^{X}\frac{1}{2}(\xi\eta^3 + \eta^2)d\xi d\eta[E_{21} + 2E_{32}, E_{23}] \\
&\quad + \frac{1}{Y}\int^{Y}\frac{1}{\eta}\int^{X}(-\xi\eta^3 + \eta^2)d\xi d\eta[E_{21} + 2E_{32}, E_{12}] \\
&= \frac{1}{Y}\int^{Y}\frac{1}{2}(\tfrac{1}{2}X^2\eta^2 + X\eta)d\eta 2(E_{33} - E_{22}) \\
&\quad + \frac{1}{Y}\int^{Y}(-\tfrac{1}{2}X^2\eta^2 + X\eta)d\eta(E_{22} - E_{11}) \\
&= \frac{1}{2}(\tfrac{1}{3}X^2Y^2 + XY)(E_{33} - E_{22}) + \frac{1}{2}(-\tfrac{1}{3}X^2Y^2 + XY)(E_{22} - E_{11}) \\
&= \frac{1}{2}(\tfrac{1}{3}X^2Y^2 + XY)E_{33} - \frac{1}{3}X^2Y^2 E_{22} - \frac{1}{2}(-\tfrac{1}{3}X^2Y^2 + XY)E_{11}.
\end{aligned}
$$

This implies

$$
\begin{aligned}
Cw^{(0)} = &\frac{1}{2}(\tfrac{1}{3}X^2Y^2 + XY)E_{33} - \frac{1}{3}X^2Y^2 E_{22} - \frac{1}{2}(-\tfrac{1}{3}X^2Y^2 + XY)E_{11} \\
&- \frac{1}{3}(E_{33} + E_{22} + E_{11}).
\end{aligned}
$$

The transformation is given by

$$
\begin{aligned}
&\frac{\alpha}{Y}\int^{Y}\frac{1}{\eta}\int^{X}\frac{1}{2}(\xi\eta^3 + \eta^2)d\xi d\eta E_{23}|_{X=1,Y=1} \\
&\quad + \frac{\alpha}{Y}\int^{Y}\frac{1}{\eta}\int^{X}(-\xi\eta^3 + \eta^2)d\xi d\eta E_{12}|_{X=1,Y=1} \\
&= \frac{\alpha}{Y}\int^{Y}\frac{1}{\eta}\frac{1}{2}(\tfrac{1}{2}X^2\eta^3 + X\eta^2)d\eta E_{23}|_{X=1,Y=1} \\
&\quad + \frac{\alpha}{Y}\int^{Y}\frac{1}{\eta}(-\tfrac{1}{2}X^2\eta^3 + X\eta^2)d\eta E_{12}|_{X=1,Y=1} \\
&= \frac{\alpha}{Y}\int^{Y}\frac{1}{2}(\tfrac{1}{2}\eta^2 + \eta)d\eta E_{23}|_{Y=1} \\
&\quad + \frac{\alpha}{Y}\int^{Y}(-\tfrac{1}{2}\eta^2 + \eta)d\eta E_{12}|_{Y=1} \\
&= \frac{\alpha}{4}(\tfrac{1}{3}Y^2 + Y)E_{23}|_{Y=1} + \frac{\alpha}{2}(-\tfrac{1}{3}Y^2 + Y)E_{12}|_{Y=1} \\
&= \frac{\alpha}{3}(E_{23} + E_{12}).
\end{aligned}
$$

We first compute the ad-action of this transformation on $v^{(2)}$. Let

$$ad(\frac{\alpha}{3}(E_{23} + E_{12}))v^{(2)}$$

$$= ad(\frac{\alpha}{3}(E_{23} + E_{12}))(E_{21} + 2E_{32} + \alpha E_{33})$$

$$= \frac{\alpha}{3}(E_{11} + E_{22} - 2E_{33} + \alpha E_{23}),$$

$$ad^2(\frac{\alpha}{3}(E_{23} + E_{12}))v^{(2)}$$

$$= \frac{\alpha^2}{9}ad(E_{23} + E_{12})(E_{11} + E_{22} - 2E_{33} + \alpha E_{23}) =$$

$$= -\frac{\alpha^2}{3}E_{23} + \frac{\alpha^3}{9}E_{13}.$$

and

$$ad^3(\frac{\alpha}{3}(E_{23} + E_{12}))v^{(2)}$$

$$= ad(\frac{\alpha}{3}(E_{23} + E_{12}))(-\frac{\alpha^2}{3}E_{23} + \frac{\alpha^3}{9}E_{13}) =$$

$$= -\frac{\alpha^3}{9}E_{13}.$$

This implies

$$v^{(3)} = \exp(ad(\frac{\alpha}{3}(E_{23} + E_{12})))v^{(2)}$$

$$= \exp(ad(\frac{\alpha}{3}E_{23}))(E_{21} + 2E_{32} + \alpha E_{33})$$

$$= E_{21} + 2E_{32} + \frac{\alpha}{3}(E_{11} + E_{22} + E_{33}) + \frac{\alpha^2}{6}E_{23} + \frac{\alpha^3}{27}E_{13}.$$

We see that $v^{(3)}$ is in not yet in normal form. The element not yet in normal form is $\frac{\alpha^2}{6}E_{23} = \frac{\alpha^2}{6}w^{(1)}$. Since we already have done the computation of $Cw^{(1)}$ we can obtain from our earlier computations that the corresponding transformation is $\frac{\alpha^2}{12}E_{13}$. We let

$$v^{(4)} = \exp(ad(\frac{\alpha^2}{12}E_{13}))v^{(3)}$$

$$= E_{21} + 2E_{32} + \frac{\alpha}{3}(E_{11} + E_{22} + E_{33}) + \frac{\alpha^2}{12}(2E_{12} + E_{23}) + \frac{\alpha^3}{27}E_{13},$$

which is the versal normal form of our original vector field v. As a check we see that the characteristic equation of the normal form (considered as a matrix) is $\lambda^2(\lambda - \alpha)$, which is equal to the characteristic equation of v itself, as it should be.

References

[CK86] A. Kelley, Cushman, R.H. and H. Kocak. Versal normal form at the Lagrange equilibrium L_4. *Journal of Differential Equations*, 64:340–374, 1986.

[LR90] R.R. London and H.P. Rogosinski. Decomposition theory in the teaching of elementary linear algebra. *American Math. Monthly*, pages 478–485, 1990.

Index

abstract, 186
action, 189, 202
algebra, 186, 187, 189, 193, 201, 202
algebraic, 186
algorithm, 186, 190, 196, 204, 205
analysis, 204
application, 195, 196, 205
apply, 186, 193, 195
associate, 201
axis, 189

bracket, 186

calculation, 188
center, 187, 189
Ch7x2x1Lem1: Lemma 2.1., 192
characteristic, 207
class, 186, 188, 191
coefficient, 203, 204
complement, 187
component, 196, 200, 205
compute, 187, 188, 190, 191, 194, 195,
    197, 200, 202, 204–206
condition, 203
coordinate, 186, 202

decomposition, 186, 187, 190, 194, 195
deformation, 204
derivative, 190
diagonal, 202
different, 190, 193
dimension, 194, 195, 200
dimensional, 188, 194, 200
direct, 200, 201
discussion, 197
divide, 200
dual, 194
dynamic, 189

eigenvalue, 192, 193, 198
eigenvector, 191–193, 196
element, 186, 187, 195, 201–204, 207

equation, 190–192, 196–198, 207
equivalence, 186, 188
equivalent, 188, 203
example, 196, 200

factor, 196
family, 187, 188
finite, 187, 188, 191, 194, 201
function, 204

general, 187, 189, 190, 195

homomorphism, 193
hypothesis, 203

ideal, 187, 203
identity, 194
image, 189, 205
imaginary, 189
independent, 194
induce, 204
induction, 203
intersection, 204
inverse, 199
invertible, 199, 203
irreducible, 194–196, 203, 204

Jacobi, 194
Jacobson-Morozov, 186, 187
Jordan, 186, 187, 190, 204

Lie, 186, 187, 189, 193, 201
linear, 186, 187, 189, 201, 202
local, 186, 203, 204

map, 199
maximal, 203
minimal, 190, 200

nilpotent, 186, 187, 189, 192, 195, 197,
    198, 203, 204
nonzero, 201
normal, 185–189, 193, 201–204, 207
number, 201

operator, 188, 191, 195, 198, 199

order, 186–190, 201

parameter, 189

point, 204

polynomial, 190, 191, 196

position, 194, 196

positive, 204

possibility, 193, 200

product, 187

projection, 192, 193, 199

property, 187, 189

prove, 199, 203

quadratic, 201

real, 204

reducible, 195

reductive, 187

representation, 185, 194–196

restriction, 204

semisimple, 187, 189–191

sequence, 188

solution, 190, 192, 197, 198, 200

solvable, 187

space, 187–189, 191, 193, 200, 201

span, 197

spectral, 194, 195

splitting, 190, 194, 195, 197, 204, 205

structure, 186

subspace, 187

theorem, 201

theory, 185–187, 194, 204

time, 193, 200

transformation, 186, 190, 204, 206, 207

triangular, 204

unique, 203

vector field, 202, 207

versal, 185, 188, 189, 201–204, 207

Painlevé Analysis and Normal Forms

L. Brenig and A. Goriely *

Abstract

This work explores various general properties of systems of nonlinear ordinary differential equations obtained by using nonlinear transformations : further reductions of Poincaré normal forms for autonomous and non-autonomous systems, Lotka-Volterra universal standard format and generalization of Painlevé singularity analysis for detecting integrable systems. These methods are progressively implemented in the computer algebra program NODES.

Contents

1. Introduction   213

2. Decoupling and integrability of ODEs through QM transformation   214

   2.1. QM transformations and representation of ODEs 214

   2.2. Exact decoupling 215

   2.3. Decoupling of Poincaré-Dulac normal forms 216

   2.4. Reduction of nonlinear systems with quasi-periodic coefficients 219

3. Lotka-Volterra universal format and applications   222

   3.1. Tranformation to Lotka-Volterra format 222

   3.2. Application: structural analysis of the Taylor expansion of the solution of (2.1) 223

*Service de Physique Statistique, Université Libre de Bruxelles, Campus de la Plaine CP231, 1050 Bruxelles, Belgium.

4. Introduction to singularity analysis   227

5. Painlevé property and Painlevé test   228

6. On the convergence of Laurent solutions   229

7. Transformation of singularities   231

   7.1. Transformation under QM transformations 232

   7.2. General recursion relations for the Laurent coefficients 232

8. Role of algebraic singularities   234

1. Introduction

The main purpose of this work is to present some general results for systems of nonlinear ODE's obtained through quasi-monomial transformations. These transformations (discovered independently by a mathematician: A. Br'uno [1]; engineers : M. Peschel and W. Mende [2]; and physicists: the authors of the present article [3, 4, 5]), provide a powerful algebraic computational scheme for the analysis of nonlinear ODE's.

Essentially, three types of results are obtained in this framework:

I. Direct decoupling and/or integrability conditions under quasi-monomial transformations (QMT) with explicit closed-form construction of the reduced ODE systems and first integrals.

II. Reduction through QM-transformation to the Lotka-Volterra standard format.

III. Extension of the Painlevé test for integrability.

- Results of type I are based on a general matrix representation of systems of ODEs with polynomial nonlinearities closely associated to the QM transformations. Decoupling and integrability conditions arise from singularities of the matrices involved in this representation [5]. Such a singularity is generic in Poincaré-Dulac normal forms. As a consequence, the dimensions of these systems are reduced after a well-defined QM transformation. This is successfully applied to systems with periodic or quasi-periodic coefficients and reduces them to autonomous systems of ODEs. These results are reported in Section 2.

- The type II results are concerned with the remarkable property that systems of nonlinear ODEs belonging to a large class can be QM-transformed into the form of Lotka-Volterra equations [2, 3]. This universal format is essentially characterized by one matrix containing all the nonlinear couplings between the dynamical variables. The general coefficients of the Taylor expansion of the solution appear to have a very simple structure in term of this matrix. This structure is completely analogous to that of partition functions of one-dimensional many-body systems in thermodynamical equilibrium. This allows for a technology transfer from the latter area to nonlinear dynamics. This is the object of Section 3. The results of category III are reported in Section 4. In this Section, we analyse the transformation of complex plane singularities of the solutions of nonlinear ODEs under the QM transformation and some special transformation of the independent variable. This leads to an extended Painlevé test for the integrability of systems of nonlinear ODEs and to a constructive way of finding first integrals. The symbolic manipulation involved by the QM framework being limited to matrix algebra, this allows for a straightforward implementation into computer algebra languages. This is realised by the program NODES (Nonlinear Differential Equations Solver) which performs algorithms generated by the above results.

2. Decoupling and integrability of ODEs through QM transformation

2.1. QM transformations and representation of ODEs

We restrict our scope to nonlinear dynamical systems which can be represented as

$$\dot{x}_i = x_i(\lambda_i + \sum_{j=1}^{m} A_{ij} \prod_{k=1}^{n} x_k^{B_{jk}}) \qquad i = 1, ..., n \qquad (2.1)$$

In equation (2.1) the dot denotes derivative with respect to the independent variable (say, the time t for instance) A and B are real or complex constant rectangular matrices and λ is a real or a complex constant vector. Most systems of physics, chemistry and biology can be cast into the form of (2.1). Along with the representation (2.1) we define the quasi-monomial (QM) transformation of the dependant variables:

$$x_i = \prod_{j=1}^{n} x_j'^{C_{ij}} \qquad i = 1, ..., n \qquad (2.2)$$

The square matrix C may be any real or complex constant invertible matrix. Hence, the set of transformations (2.2) is a nonlinear representation of $GL(n, C)$.

Now, the main interest of these associated representations and transformations lies in the fact that form (2.1) is invariant under the above transformations. Indeed, after performing one of these, one gets

$$\dot{x}_i' = x_i'(\lambda_i' + \sum_{j=1}^{m} A_{ij}' \prod_{k=1}^{n} x_k'^{B_{jk}'}) \qquad i = 1, ..., n \qquad (2.3)$$

with

$$\begin{cases} \lambda' = C^{-1}.\lambda \\ A' = C^{-1}.A \\ B' = B.C \end{cases} \qquad (2.4)$$

This is to be compared with the form-invariance of linear ODEs

$$\dot{x}_i = \sum_{j=1}^{n} L_{ij} x_j \qquad i = 1, ..., n \qquad (2.5)$$

under linear transformations

$$x'_i = \sum_{j=1}^{n} T_{ij} x_j \qquad i = 1, ..., n \qquad (2.6)$$

which transforms into

$$\dot{x}'_i = \sum_{j=1}^{n} L'_{ij} x'_j \qquad i = 1, ..., n \qquad (2.7)$$

with

$$L' = T^{-1}.L.T \qquad i = 1, ..., n \qquad (2.8)$$

In the same way as for linear systems which can be transformed into simpler forms (Jordan, diagonal) by an appropriate linear transformation, we may look for particular QM-transformations such that (2.3) is simpler than (2.1). This is generally possible and leads to the Lotka-Volterra form which is discussed in the next Section. Moreover, in cases of singularity of matrices A and B, we get exact decoupling and/or first integrals as we now present.

2.2.   Exact decoupling

Proposition 2.1 *If in equation (2.1) the $m \times n$ matrix B is of rank r, then (2.1) can be transformed by an appropriate QM-transformation into an r-dimensional nonlinear system and an $(n-r)$-dimensional linear system. The r-dimensional system is decoupled from the $(n-r)$-dimensional one.*

Proof : if B is of rank r, it admits $(n-r)$ vectors $\phi^{(\alpha)}$, $\alpha = r+1, ..., n$, such that $B.\phi^{(\alpha)} = 0$.

Let us apply to equation (2.1) a QM transformation (2.2) with the following particular matrix C:

$$C = \begin{pmatrix} I_{r \times r} & & & \\ & \phi^{(r+1)} & ... & \phi^{(n)} \\ O_{(n-r) \times r} & & & \end{pmatrix} \qquad (2.9)$$

where $I_{r \times r}$ is an $r \times r$ identity matrix, $O_{(n-r) \times r}$ is a null matrix and the next columns are the $(n-r)$ null-eigenvectors of B.

The transformed equations are written in the form (2.3) - (2.4) with

$$B' = B.C = \begin{pmatrix} B_{mxr} & O_{mx(n-r)} \end{pmatrix}$$ (2.10)

where B_{mxr} is the restriction to the first r columns of the matrix B. The structure (2.10) of B' clearly implies that the product in equation (2.3) is in fact restricted to the first r factors, the $(n-r)$ remaining factors being equal to one. Hence, in this case the equations for the first r variables are given by

$$\dot{x}'_i = x'_i(\lambda'_i + \sum_{j=1}^{m} A'_{ij} \prod_{k=1}^{r} x'^{B'_{jk}}_k)   \qquad i = 1, ..., r$$ (2.11)

which is a closed system decoupled from the $(n-r)$ remaining equations :

$$\dot{x}'_i = x'_i(\lambda'_i + \sum_{j=1}^{m} A'_{ij} \prod_{k=1}^{r} x'^{B'_{jk}}_k)   \qquad i = r+1, ..., n$$

which appear as linear equations in $x'_{r+1}, ..., x'_n$ with time-dependent coefficients depending on the dynamics of the first r variables.

Thus, when B is of rank $r < n$, the relevant nonlinear dynamics is reduced from n dimensions to r. As a consequence, the non-trivial properties as the existence of periodic or chaotic attractors are living in a r-dimensional phase space.

We have developed rather lengthily the consequences of the singularity of matrix B since this property is generic to Poincaré-Dulac normal forms as shown in the next subsection. However, a similar property appears when the matrix A is singular of rank r. In the same way, it is shown [5] that this leads to the existence of $(n-r)$ first integrals obeying linear equations and which can be obtained in closed form. This is a property dual to the previous one which is related to a more general duality between A and B.

2.3. Decoupling of Poincaré-Dulac normal forms

Let us consider a system of ODEs with diagonalized linear operator:

$$\dot{x}_i = \lambda_i x_i + \sum_{(m_1,...,m_n)} a_i(m_1, ..., m_n)x_1^{m_1}...x_n^{m_n}   \qquad i = 1, ..., n$$ (2.12)

with $m = m_1 + m_2 + ... + m_n$ ($m_i \in Z \ \forall \ i$ and $m > 1$) and let us apply a Poincaré transformation to the dependent variables defined as follows:

$$x_i = x_i' + \sum_{(m_1,...,m_n)} b_i(m_1, ..., m_n) x_1^{m_1} ... x_n^{m_n} \qquad i = 1, ..., n; \quad m > 1$$

(2.13)

which, in the vicinity of the origin, is a near-identity transformation.

The Poincaré-Dulac theorem [6] states under rather broad conditions on equation (2.12) that there exists a transformation of type (2.13) leading to

$$\dot{x}_i' = \lambda_i x_i' + \sum_{(m_1,...,m_n)} c_i(m_1, ..., m_n) x_1'^{m_1} ... x_n'^{m_n} \qquad i = 1, ..., n \quad (2.14)$$

and such that

$$c_i(m_1, ... m_n) = \begin{cases} 0 & \text{for } m_1, ..., m_n \text{ such that } \lambda_1 m_1 + ... + \lambda_n m_n \neq \lambda_i \\ c_i & \text{otherwise} \end{cases}$$

(2.15)

The condition $\lambda_1 m_1 + \lambda_2 m_2 + ..., \lambda_n m_n = \lambda_i$ is called a *resonance* condition and the corresponding monomial $x_1^{m_1} ... x_n^{m_n}$ is a *resonant monomial*.

Obviously (from 2.15), any equation for which no resonance condition is fulfilled will turn into a linear equation after the transformation (2.13):

$$\dot{x}_i' = \lambda_i x_i' \qquad i = 1, ..., n \qquad (2.16)$$

The equations containing resonances remain nonlinear but with only resonant monomials as nonlinearities. Equations (2.14) with conditions (2.15) are called *normal forms* and the corresponding transformation (2.13) a *Poincaré transformation*.

Our scope here is concerned neither with the computation of the coefficients nor with the question of the convergence or summability of the associated series. These questions are developed in other papers of this volume (see for example, the papers by Della Dora and Stolovitch and by Sibuya).

We restrict our interest to the normal form equations themselves, assuming that they exist. We now show that these equations are generically reducible from n to $(n-1)$ dimensions through a QM transformation (for a similar result, see also [1]).

To show this, let us write down the equations (2.12) in the form (2.1):

$$\dot{x}_i = x_i(\lambda_i + \sum_{j=1}^{m} A_{ij} \prod_{k=1}^{n} x_k^{B_{jk}}) \qquad\qquad i = 1,...,n \qquad (2.17)$$

where m may be infinite.

The resonance condition (2.15) is written as

$$\sum_{k=1}^{n} B_{jk}\lambda_k = 0 \qquad (2.18)$$

for a given $j \in \{1,...,m\}$.

The corresponding resonant monomial appears in the i-th equation of (2.17) provided that

$$A_{ij} \neq 0 \qquad (2.19)$$

Now if (2.17) is already supposed to be a normal form, this means that all the monomials in (2.17) are resonant. That is, equation (2.18) is valid for any $j = 1,...,m$. Thus, the vector λ belongs to the kernel of the matrix B which is at most of rank $n-1$. This corresponds to a case where proposition 2.1 applies. Hence, we may state the

Proposition 2.2 *Any n-dimensional Poincaré normal form may be transformed into an (n-1)-dimensional system of ODEs by an appropriate QM transformation.*

Moreover, that transformation is generated by the matrix

$$C = \begin{pmatrix} & & \lambda_1 \\ & I_{(n-1)\times(n-1)} & \vdots \\ & & \lambda_n \end{pmatrix} \qquad (2.20)$$

Proof :

Let us explicitly perform this transformation on the normal form (2.17):

$$x_i = \prod_{j=1}^{n} x_j'^{C_{ij}} \qquad i = 1, ..., n \qquad (2.21)$$

We get for the new variables

$$\dot{x}_i' = x_i' \sum_{j=1}^{m} (A_{ij} - \frac{\lambda_i}{\lambda_n} A_{nj}) \prod_{k=1}^{n-1} x_k'^{B_{jk}} \qquad i = 1, ..., n-1 \qquad (2.22)$$

and

$$\dot{x}_n' = x_n'(1 + \sum_{j=1}^{m} \frac{1}{\lambda_n} A_{nj} \prod_{k=1}^{n-1} x_k'^{B_{jk}}) \qquad (2.23)$$

where we have used relations (2.4) to find the new matrix A' and vector λ'. Equation (2.22) is an $(n-1)$-dimensional nonlinear system and (2.23) is a 1-dimensional linear equation with time-dependent coefficients.

As a consequence, a 2-dimensional normal form can always be reduced to a 1-dimensional nonlinear autonomous equation which in turn can be solved by quadrature. Such a property appears clearly in many examples among which the case of the perturbed harmonic oscillator, discussed in the article by Meyer in the present volume, is a good illustration.

Another important application of proposition 2.2 is the reduction of normal forms with quasi-periodic coefficients to autonomous systems of the same dimension as we show in the next subsection.

2.4. Reduction of nonlinear systems with quasi-periodic coefficients

To grasp the main idea of this result, let us show how it works on a particular case [7]. Let us consider the equation

$$\ddot{x} = f(x, \dot{x}, g(t)) \qquad (2.24)$$

where $g(t)$ is a function of period $\frac{2\pi}{\omega}$. An example would be the periodically forced Van der Pol oscillator.

We first introduce the new variables x.

$$\begin{cases} x_1 = \dot{x} \\ x_2 = x \end{cases} \tag{2.25}$$

In the next step, we expand $g(t)$ in Fourier series,

$$g(t) = \sum_{l=1}^{\infty} (u_l \sin(l\omega t) + v_l \cos(l\omega t)) \tag{2.26}$$

and define the variables

$$\begin{cases} x_l^{(u)} = u_l \sin(l\omega t) \\ \bar{x}_l^{(u)} = u_l \cos(l\omega t) \\ x_l^{(v)} = v_l \cos(l\omega t) \\ \bar{x}_l^{(v)} = v_l \sin(l\omega t) \end{cases} \tag{2.27}$$

In terms of these variables the equation (2.24) becomes

$$\begin{cases} \dot{x}_1 = f(x_1, x_2, x_1^{(u)}, \bar{x}_1^{(u)}, x_1^{(v)}, \bar{x}_1^{(v)}, \ldots) \\ \dot{x}_2 = x_1 \\ \dot{x}_l^{(u)} = l\omega \bar{x}_l^{(u)} \\ \dot{\bar{x}}_l^{(u)} = -l\omega x_l^{(u)} \\ \dot{x}_l^{(v)} = -l\omega \bar{x}_l^{(v)} \\ \dot{\bar{x}}_l^{(v)} = l\omega x_l^{(v)} \end{cases} \tag{2.28}$$

and where the initial conditions are determined by (2.27) at time $t = 0$. Equation (2.28) is an infinite-dimensional system ($l = 1, \ldots, \infty$).

Let us assume that $f(x_1, x_2, g(t))$ is of the form

$$f(x_1, x_2, g(t)) = ax_1 + bx_2 + h(x_1, x_2, g(t)) \tag{2.29}$$

that is, it possesses a linear part in x_1, x_2 . We first diagonalize the linear operator of system (2.28). This is readily done by separately diagonalizing the 2×2 linear operator on x_1, x_2 and the sequence of 2×2 linear operators acting on the variables $\{x_l^{(u)}, \bar{x}_l^{(u)}, x_l^{(v)}, \bar{x}_l^{(v)}\}$. In terms of the new variables we get

$$\begin{cases} \dot{y}_1 = \lambda_1 y_1 + N_1(y_1, y_2, y_3, \ldots) \\ \dot{y}_2 = \lambda_2 y_2 + N_2(y_1, y_2, y_3, \ldots) \\ \dot{y}_k = \lambda_k y_k \qquad\qquad k = 3, \ldots, \infty \end{cases} \tag{2.30}$$

Let us now consider an arbitrary truncation of order K of the above Fourier series ($l \leq K$). This yields a finite system of ODEs of dimension $N = 4K + 2$. We perform the transformation of type (2.13) leading to the following normal form.

$$
\begin{cases}
\dot{z}_1 = \lambda_1 z_1 + \sum_{m_1,...,m_n} g_1(m_1,...,m_n)z_1^{m_1}...z_N^{m_N} \\
\dot{z}_2 = \lambda_2 z_2 + \sum_{m_1,...,m_n} g_2(m_1,...,m_n)z_1^{m_1}...z_N^{m_N} \\
\dot{z}_k = \lambda_k z_k \qquad\qquad\qquad k = 3,...,N
\end{cases}
\tag{2.31}
$$

where $N = 4K + 2$ and the r.h.s. of the first two equations contain only resonant monomials. Remark that the remaining equations are unchanged by construction of the Poincaré transform.

Let us now apply to system (2.31) a QM transformation (2.2) with a $N \times N$ matrix C of form (2.20). We get

$$
\begin{cases}
\dot{z}_1' = \sum_{m_1,...,m_{N-1}} g_1'(m_1,...,m_{N-1})z_1'^{m_1}...z_{N-1}'^{m_{N-1}} \\
\dot{z}_2' = \sum_{m_1,...,m_{N-1}} g_2'(m_1,...,m_{N-1})z_1'^{m_1}...z_{N-1}'^{m_{N-1}} \\
\dot{z}_k' = 0 \qquad\qquad\qquad k = 3,...,N-1 \\
\dot{z}_N' = z_N'
\end{cases}
\tag{2.32}
$$

where g_1', g_2' are calculated by using the equation (2.22). Thus the variables z_3' to z_N' are constants of motion and are determined by the initial condition, whereas Z_N' is an exponentially time-dependant first integral also determined by the initial conditions. Combining these results along with the decoupling property of the first $N - 1$ variables from the N-th one, we obtain a set of 2 coupled autonomous equations for z_1' and z_2' :

$$
\begin{cases}
\dot{z}_1' = \sum_{m_1,m_2} c_1'(m_1,m_2)z_1'^{m_1}z_2'^{m_2} \\
\dot{z}_2' = \sum_{m_1,m_2} c_2'(m_1,m_2)z_1'^{m_1}z_2'^{m_2}
\end{cases}
\tag{2.33}
$$

Since this result is independent of the choice of the truncation K of Fourier series (2.26), it can be extended to the full system (2.24) provided the series involved in that step converge. We have thus shown that a 2-dimensional differential system like (2.24) with periodic coefficients can be reduced to a 2-dimensional autonomous system by performing a Poincaré transformation followed by an appropriate QM transformation. The only memory of the non- autonomous character of (2.24) is the existence of a time-dependent first integral. The constants of motion mentioned above are related to the coefficients of the Fourier series (2.26).

Obviously, this result also holds for a quasi-periodic function $g(t)$ depending on several frequencies $\omega_1, ..., \omega_s$. This is shown by expanding $g(t)$ in s-dimensional Fourier series and by truncating at an arbitrary mode: the reasoning then follows the same steps.

Moreover, the above proof can be extended to non-autonomous of dimension higher than two without any difficulty. Hence, we can state the

Proposition 2.3 *Any n-dimensional system of ODEs with quasi-periodic right hand side and diagonalizable linear part can be transformed into an autonomous system of the same dimension provided its Poincaré normal form exists.*

This result is a constructive one since the QM transformation is known in closed form. However, generally the Poincaré normalizing tranformation is only obtained as a perturbative expansion whose convergence has to be investigated. Several theorems on the convergence of normal forms are available but many situations fall outside the conditions of these theorems and the convergence must be checked case by case.

3. Lotka-Volterra universal format and applications

3.1. Tranformation to Lotka-Volterra format

A glance at equations (2.3) and (2.4) shows that in the particular case where $m = n$, a QM tranformation with $C = B^{-1}$ (provided $\det(B) \neq 0$ exists) transforms (2.1) into

$$\dot{x}_i' = x_i'(\lambda_i' + \sum_{j=1}^{m} A_{ij}' x_j') \tag{3.1}$$

$$\begin{cases} \lambda' = B.\lambda \\ A' = B.A \end{cases} \tag{3.2}$$

This is due to the fact that in this transformation $B' = I$.

Equation (3.1) is of the form of the famous Lotka-Volterra equation introduced first in theoretical ecology [8].

The main characteristic of this class of systems is that it presents the lowest nonlinearity (quadratic) and the simplest, since the most general quadratic nonlinear system would involve a term $N_{ijk}x_j x_k$, still preserving the possibility of complex behaviour like chaos [9]. The degeneracy of the tensor N_{ijk} into the matrix A'_{ij} cannot be extended further. Indeed, let us consider the case where the matrix A' in (3.1) would itself degenerate into a tensor product of two vectors. Then the system would become completely integrable since in this case the rank of A' is 1 and one would find a QM transformation providing $(n-1)$ first integrals. Thus, chaos would be impossible in such systems.

In fact, even when m is different from n, it has been shown that the result (3.1) holds [2, 3, 4]. However, the number of dimension of the resulting Lotka-Volterra equations is now m instead of n with the same transformation rules (3.2) for the coefficients. This result is obtained by a straightforward reasoning implying an embedding into m dimensions of the former n dimensional system (2.1). It is true provided the rank of B is maximum. If the rank of B is smaller, say r, we know that a particular QM transformation would decouple the n-dimensional system into an r-dimensional system whose matrix B is of maximal rank. Then, this last system, in turn, may be transformed into the Lotka-Volterra format. Thus, the Lotka-Volterra format is always obtained for systems which can be written as (2.1).

3.2.   Application: structural analysis of the Taylor expansion of the solution of (2.1)

Let us perform the following steps starting with equation (2.1):

Step 1: Perform the following linear transformation:

$$x_i = e^{\lambda_i t} y_i \qquad (3.3)$$

This gives

$$\dot{y}_i = y_i \sum_{j=1}^{m} A_{ij} \prod_{k=1}^{n} y_k^{B_{jk}} e^{\gamma_j t} \qquad (3.4)$$

with

$$\gamma_j = \sum_{k=1}^{n} B_{jk} \lambda_k \qquad (3.5)$$

Step 2: Define a new dependant variable

$$y_{n+1}(t) = e^t \qquad (3.6)$$

Step 3: Write the system (3.4) in the form

$$\dot{y_i} = y_i \sum_{j=1}^{m+1} \tilde{A}_{ij} \prod_{k=1}^{n+1} y_k^{\tilde{B}_{jk}} \qquad\qquad i = 1, ..., n+1 \qquad\qquad (3.7)$$

with

$$y_{n+1}(t=0) = 1$$

and

$$\tilde{B} = \begin{pmatrix} & & & \gamma_1 \\ & B_{m \times n} & & \vdots \\ & & & \vdots \\ & & & \gamma_m \\ 0 & \cdots & 0 & 0 \end{pmatrix} \qquad\qquad (3.8a)$$

$$\tilde{A} = \begin{pmatrix} & & & 0 \\ & A_{n \times m} & & \vdots \\ & & & \vdots \\ & & & 0 \\ 0 & \cdots & 0 & 1 \end{pmatrix} \qquad\qquad (3.8b)$$

Step 4: Perform on system (3.7) the QM transformation leading to the Lotka-Volterra format. This is possible if $\operatorname{rank}(B) = n$, since in this case $\operatorname{rank}(\tilde{B}) = n+1$.

The transformed equations reads:

$$\dot{z_i} = z_i \sum_{j=1}^{m+1} (\tilde{B}.\tilde{A})_{ij} z_j \qquad\qquad i = 1, ..., m+1 \qquad\qquad (3.9)$$

and where

$$z_i = \prod_{j=1}^{n+1} y_j^{\tilde{B}_{ij}} \qquad\qquad i = 1, ..., m+1 \qquad\qquad (3.10)$$

Step 5: Rescale the variables z_i by

$$u_i = \frac{z_i}{z_i(t=0)} \qquad\qquad (3.11)$$

This excludes initial conditions on the coordinate planes.

The equations for u_i are now:

$$\dot{u}_i = u_i \sum_{j=1}^{m+1} M_{ij} u_j \qquad i = 1, ..., m+1 \qquad (3.12)$$

with

$$M_{ij} = (\tilde{B}.\tilde{A})_{ij} z_j (t = 0) \qquad (3.13)$$

Step 6: Construct the Taylor expansion for $u_i(t)$

$$u_i(t) = \sum_{k=0}^{\infty} \frac{t^k}{k!} c_i(k) \qquad (3.14)$$

where

$$c_i(k) = (\frac{d^k}{dt^k} u_i(t))_{t=0} \qquad (3.15)$$

The $c_i(k)$ are obtained by iterating (3.12), taking into account that $u_i(0) = 1$ for all $i = 1, ..., m+1$. This procedures gives

$$c_i(k) = \sum_{i_1=1}^{m+1} ... \sum_{i_k=1}^{m+1} M_{ii_1}(M_{ii_2} + M_{i_1 i_2})(M_{ii_3} + M_{i_1 i_3} + M_{i_2 i_3})...$$

$$...(M_{ii_k} + M_{i_1 i_k} + ... + M_{i_{k-1} i_k})$$

$$(3.16)$$

This is the main result. It provides a deep insight into the structure of the Taylor series for the solution of systems of nonlinear ODEs. It is interesting to compare this expression to the coefficients of the Taylor expansion for the solution of a system of linear ODEs:

$$\dot{x}_i = \sum_{j=1}^{m+1} M_{ij} x_j, \quad i = 1, ..., m+1 \qquad (3.17)$$

We get in this case

$$
\begin{aligned}
c_i(k) &= \sum_{i_1=1}^{m+1} \cdots \sum_{i_k=1}^{m+1} (M_{ii_1} M_{i_1 i_2} \ldots M_{i_{k-1} i_k}) \\
&= \sum_{i_k=1}^{m+1} (M^k)_{ii_k}
\end{aligned}
$$

$$(3.18)$$

That is, this result inserted in the Taylor series for $x_i(t)$ provides the exponential of tM.

In fact, the expression (3.18) corresponds to only one term among $k!$ in the expansion of the products in (3.16).

Graphs may be introduced describing each of these $k!$ terms. The elementary building block of these graphs is given by

$$= M_{ij}$$

$$(3.19)$$

Thus we have

$$= M_{ii_1} M_{i_1 i_2} \ldots M_{i_{k-1} i_k}$$

$$(3.20)$$

representing the terms in (3.18).

Another example is

$$= M_{ii_1} M_{i_1 i_2} \ldots M_{i_{k-1} i_k}$$

$$(3.21)$$

representing the term in (3.16) obtained by taking as factors the first terms of all the parentheses.

The graphs are trees rooted at vertex i. Partial summation of the Taylor series may be achieved by summing subseries corresponding to graphs having the same topology (for example the already mentioned terms of form (3.18), corresponding to topology (3.20), may be summed and give the exponential of M). Such graphs are similar to those introduced by R. Grossman in his algorithm for computing approximate solutions to ODEs [10].

Furthermore, the structure (3.16) is strongly reminiscent of the form of a partition function in equilibrium statistical mechanics. Indeed for an N-body quantum system whose levels are characterized by the quantum numbers $i_1, ..., i_N$ such that $i_j = 1, ..., n$ $\forall$ $j = 1, ..., N$ and whose energy-eigenvalues are given by a function $\epsilon(i_1, ..., i_N)$ on these levels, the partition function is given by

$$Z = \sum_{i_1=1}^{m} ... \sum_{i_k=1}^{m} e^{-\beta \epsilon(i_1, ..., i_N)} \tag{3.22}$$

where β is essentially the inverse of the temperature, i.e. a control parameter.

This is the central object in the calculation of the thermodynamic quantities like specific heat or pressure from the microscopic level.

It appears that (3.16) may be related to such a partition function. This allows for a technology transfer from equilibrium statistical mechanics to nonlinear dynamics. In particular, the birfucation of the dynamics under parameter changes would be related to equilibrium phase changes (like solid-liquid or ferromagnetic-paramagnetic) for which a powerful theory based on the renormalization group exists [10]. These ideas are under study and will be published elsewhere.

4.   Introduction to singularity analysis

In the field of ODEs there exist many different approaches, each one providing different answers to different questions. One of the most fundamental questions lies in the choice of the equations that should be analysed. For a mathematician, a class of DE is considered (e.g. ODE with analytic vector fields, linear partial differential equations, hyperbolic systems...) and general properties such as existence of solutions, boundedness of solutions in a given functional space are sought. Another important classical question is the definition of new functions, new transcendents defined via the solution of a DE. [1]

---

[1] NOTE the first important case being the definition of the Log by Napier via the "unphysical" equation of motion suggested first by Galileo with speed proportional to space.

This implies that the solution is well-defined and single valued. These global properties of solutions in the complex plane of the independent variables led Poincaré first and Painlevé to consider all first and second order equations for which the general solution is single valued and finally led to the definition of six new functions known as Painlevé transcendents [11]. Another fruitful approach is the physicist's approach. In this case, some equations are deduced from general laws and first principles taking into account local interactions of particles, fluid elements or more generally the motion of point masses under the action of force fields. These "god-given" equations should be in a certain way "solved", meaning that any information on the global character of the solutions is of fundamental importance. Many different ways of effectively solving equations have been considered, originally, by Euler, Bernoulli, Riccatti and others. The zenith of these constructive methods was reached in the field of classical mechanics with the work of S. Kowaleskaya on the spinning top [12]. It was realized on this problem that all analytic solutions had been found but also and more important that there was no other solution.

5.   Painlevé property and Painlevé test

The class of systems (referred as to the QM class) we are interested in, is of the form

$$\begin{cases} \dot{x}_i = f_i(x_1, ..., x_n) \\ x_i(0) = x_0 \end{cases} \tag{5.1}$$

where f_i is a finite sum of terms of the form $x_1^{k_1}...x_n^{k_n}$ with $k_i \in Q$.[2]

As we are interested in the behaviour of the solutions near the singularities in the complex plane of the independent variables, the "time" t is considered to be complex.

Definition 5.1 *A movable singularity is a singularity whose location depends on the initial conditions.*

Definition 5.2 *A critical singularity of a function is a singularity around which the valuations are permuted.*

Definition 5.3 *The system (5.1) has the Painlevé property (PP) if the general solution does not exhibit any movable critical singularity.*

---

[2]Let us point out that the singularity analysis does not make sense for general analytic functions since the behaviour near the singularity is determined by the highest power of f_i as shown later on.

There has been much interest in the last decade in the singularity analysis in relationship with integrability. It was found that for many integrable PDEs (integrable via the Inverse Scattering Techniques), the similarity reductions of these PDEs lead to ODEs exhibiting the PP. This leads to the conjecture that every similarity reduction of integrable PDE has the PP. Even if a full proof of this statement is still lacking, a partial proof or theoretical justification can be found in [13] On the other hand the PP for dynamical systems and Hamiltonian vector fields was largely investigated using the so-called Painlevé test for ODEs. The Painlevé test is an algorithmic procedure for determining necessary conditions for the PP. It has been shown that this test is not sufficient in many cases and that the hope of a finite algorithmic and systematic procedure for detecting the PP remains elusive.

6.   On the convergence of Laurent solutions

Definition 6.1 *The system (5.1) is weight-homogeneous if there exists a similarity transformation $x_i \rightarrow \alpha x_i^{g_i}$ such that*

$$f_i(\alpha^{g_1} x_1, ..., \alpha^{g_n} x_n) = \alpha^{w_i} f_i(x_1, ..., x_n)$$

It appears that the similarity transformations and the weight-homogeneous property of systems are the building blocks in the singularity analysis as they determine the behaviour of the solutions at the singularity. They are therefore a typical feature of nonlinear systems.

Theorem 6.1 (Adler and Van Moerbeke, 1989) *The formal Laurent solutions*

$$x_i(t) = \alpha_i \tau_i^p \sum_{j=0}^{\infty} a_{ij} \tau^j \qquad \tau = t - t_0 \qquad (6.1)$$

of a weight-homogeneous system (5.1) are convergent series.

This result can be easily generalized to arbitrary vector fields belonging to the QM class.

Theorem 6.2 *The formal Laurent solutions of a system (1.1) are convergent series.*

Proof :   Let us consider a QM system (1.1)

$$\dot{x}_i = x_i \sum_{j=1}^{m} A_{ij} \prod_{k=1}^{n} x_k^{B_{jk}}$$

(6.2)

admitting Laurent solutions (6.1). We first assume that B is of maximal rank. As shown in Section 3, (6.2) can be transformed into an LV system without linear terms. Such an LV system is homogeneous, the Laurent solutions are thus convergent series. We now show that these series are mapped via a QMT on convergent series solutions of the original problem. As introduced into Section 3, the mapping of system (6.2) onto an LV system is made through the QM transformation:

$$x_i' = \prod_{j=1}^{n} x_j^{B_{ij}}$$

(6.3)

In the new variables, system (6.2) reads

$$\dot{x}_i' = x_i' \sum_{j=1}^{m} M_{ij} x_j'$$

(6.4)

with $M = B.A$.

This LV system is a weight-homogeneous system of weight $\nu_i' = 1 \ \forall i$. By theorem 6.1, it is known that

$$x_i'(t) = \alpha_i' \tau^{p_i'} (T_i'(\tau))$$

(6.5)

where $T_i'(\tau)$ is some convergent Taylor series in t, is a convergent solution of system (6.4). Given this convergent series, the inverse transformation univocally defines a convergent series $x = x(\tau)$ which is solution of the original system. The inverse transformation can be written

$$x_j = \exp(\sum_k (x_k' \log(C_{jk}) + 2l\pi i C_{jk}))$$

(6.6)

with $l \in Z$ and where $C = B^{-1}$.

Since (6.2) admits Laurent solutions, the multiplicity factor must simplify, that is

$$\sum_j C_{ij} = p_i \tag{6.7}$$

with $p_i \in Z$ by hypothesis. The inverse transformation now reads

$$x_j = \exp(\sum_k (x'_k \log(C_{jk}))) \tag{6.8}$$

The Laurent solutions $x'_i = x'_i(t)$ convergent in a punctured domain around $t = t_0$ are mapped by (6.8) onto convergent Laurent solutions

$$x_i = \alpha_i \tau^{p_i}(T_i(\tau)) \tag{6.9}$$

which proves the result in the case where B is of maximal rank.

If B is not of maximal rank, it has been shown in Section 2 that the system is in a certain sense degenerate and can be reduced to an r-dimensional system (where $r =$rank(B)) on which the previous proof can be applied.

7.   Transformation of singularities

In this section we use the explicit closed form of the vector field transformations introduced in Section 2 to study the transformation of singularities in the complex plane by following the transformation of the Laurent series representing the behaviour of the solution near a singularity. Let us consider a vector field belonging to the QM class

$$\dot{x} = f(x) \tag{7.1}$$

and let us assume that the extension of (7.1) in the complex plane of time $(t \in R \rightarrow t \in C)$ admits around a movable singularity a power expansion as a solution:

$$x = \alpha \tau^p \sum_{i=0}^{\infty} a_i \tau^i \tag{7.2}$$

with $\tau = (t - t_0)$, $a_0 = 1$ and α, p constant vectors.

We now show that the coefficients a_i are defined via a recursion relation of the form

$$Q.a_i = ia_i + P(a_1, ..., a_{i-1}) \tag{7.3}$$

Let $R = \{-1, r_2, ..., r_n\}$ be the set of resonances. These are by definition the eigenvalues of the recursion operator Q (and should not be confused with Poincaré resonances introduced above).

The Painlevé test is an algorithmic procedure which proceeds as follows. First, find all possible singular behaviours of the solutions near a singularity. This is achieved by substituing $x_i = \alpha_i \tau^{p_i}$ in (7.1) and keeping the leading order terms as $\tau \to 0$. This gives a collection of vectors (α, p). Second, for each pair (α, p) (defining a *branch*) compute the eigenvalues of matrix Q (cf (7.3)) defining the recursion relation. Third, check the formal existence of the Laurent solutions (7.2) for (7.1) for each branch. To do so, we notice that each resonance $i \in R$ in (7.3) provides a compatibility condition for the solvability of a linear system (Fredholm alternative). If these compatibility conditions are satisfied, the branch (α, p) defines a Laurent expansions since all coefficients can be found using (7.3). Finally, if all branches define Laurent expansions, the system passes the Painlevé test. If for any branch, either p or R is not integer or a compatibility condition is not satisfied, the system fails to pass the test.

7.1.   Transformation under QM transformations

It can be shown [15] that under a QMT (2.2) the parameters characterizing the Laurent expansion (α, p, R) are mapped onto

$$\begin{cases} \alpha & \to \alpha^C \\ p & \to C^{-1}.p \\ R & \to R \end{cases} \tag{7.4}$$

It is clear that using the QM transformation it is always possible to find a representation in which the prefactor $\alpha \tau^p$ can be mapped onto a single pole $p = -1$ by choosing for instance the Lotka-Volterra representation. The possibility of allowing a non-integer coefficient p is due to an improper choice of the variables in which the analysis is performed.

7.2.   General recursion relations for the Laurent coefficients

The LV canonical form introduced in [3] seems to be the ideal framework to study the Painlevé test since it has the lowest homogeneous nonlinearity (quadratic) and therefore the leading behaviour is only characterized by pure poles. Moreover, for the sake of clarity we will focus our study on the most dominant balance, that is the Laurent expansion solution of 3.1 with $p_i = -1 \ \forall \ i$ and $\alpha_i \neq 0 \ \forall \ i$ The general result for an arbitrary balance can be found in [15].

The LV system reads

$$\dot{x}_i = x_i \sum_{j=1}^{n} M_{ij} x_j \tag{7.5}$$

and the Laurent series solution of (7.1) that we consider is

$$\begin{cases} x_i = \alpha \tau^{-1} \sum a_{ij} \tau^j \\ a_0 = 1 \end{cases} \tag{7.6}$$

with $\alpha = -M^{-1}.(1)$ where $(1)_i = 1 \ \forall \ i$.

It is easy to show that the resonance set is given by the eigenvalues of the following recursion operator:

$$Q_{ij} = M_{ij} \alpha_j \tag{7.7}$$

Once the matrix Q is known, the recursion relation for the coefficient of the Laurent series can be computed.

$$\begin{aligned} Q.a_k &= P_k \\ &= ka_k - \sum_{i=1}^{k-1} a_i P_{k-i} \end{aligned} \tag{7.8}$$

with $P_0 = 0$.

The recursion relation for P can be explicitly inverted to obtain

$$\begin{cases} P_k = \sum_{\{i_j\}} C(\{i_j\}) \prod_{j=1}^{n} a_j^{i_j} \\ C(\{i_j\}) = \frac{\beta!(-)^\beta}{\prod_j i_j!} \end{cases} \tag{7.9}$$

with $\beta = -1 + \sum i_j$ where the indices $i_j \in N$ are such that $\sum_{j=1}^{n} i_j j = n$.

This form of the recursion relation is convenient to prove the

Proposition 7.1 *If the set of resonances of the Laurent series is such that*

$$r_i \neq \sum n_j r_j \qquad\qquad \forall \ n_j \in N \ \forall \ r_i \in R$$

then the compatibility conditions are satisfied.

This means that knowing the set of resonances for a given branch is enough for ensuring the formal existence of the series whose convergence can then be ensured by theorem 6.2

Proof :

Since $r_i \neq \sum n_j r_j$ the P_i reduces to

$$P_{r_i} = r_i a_{r_i}$$

The compatibility conditions

$$Q.a_{r_i} = r_i a_{r_i}$$

are automatically satisfied by taking a_{r_i} as eigenvectors of the matrix Q of eigenvalues r_i.

8.   Role of algebraic singularities

Definition 8.1 *A system is complex analytically integrable if there exists a complete set of analytic integrals of motion*

The Painlevé test in the given form does not allow algebraic singularities since they are critical. However it seems that certain types of algebraic singularities can be related to the complex analytical integrability. Indeed, it was found that some 2-dimensional Hamiltonians such as

$$H = \frac{1}{2}(p_x^2 + p_y^2) + 64y^6 + 80x^2y^4 + 24x^4y^2 + x^6 \tag{8.1}$$

with second invariant

$$I = -yp_x^2 + xp_xp_y + 32x^2y^5 + 32x^4y^3 + 6x^6y \tag{8.2}$$

only exhibit algebraic singularities. This leads to

Definition 8.2 *A system (1.1) is of weak-Painlevé type (or satisfies the weak-Painlevé test) if all solutions can be expanded in Puiseux series:*

$$x_i = \tau^{p_i} \sum_j a_{ij} \tau^{j/m_i} \tag{8.3}$$

with $p_i = g_i/m_i$.

We now show how the weak-Painlevé systems can be connected to Painlevé type systems via an appropriate nonlinear transformation. We also show that in general, weak-Painlevé systems cannot be connected to complex analytical integrability due to the existence of chaotic systems of the weak-Painlevé type. For that purpose, we introduce a new class of nonlinear transformations acting on the QM representation, the so-called *New-Time Transformations* (NTT):

$$dt = (\prod_{i=1}^{n} x_i^{\beta_i}) d\tilde{t} \tag{8.4}$$

with $\beta_i \in R \; \forall \, i$

Under an NTT, a system written in the QM representation is mapped onto:

$$\frac{dx_i}{d\tilde{t}} = x_i \sum_{j=1}^{m} A_{ij} \prod_{k=1}^{n} x_k^{\tilde{B}_{jk}} \tag{8.5}$$

with

$$\tilde{B}_{ij} = B_{ij} + \beta_j \tag{8.6}$$

We are now interested in the transformation of the Laurent series solutions under an NTT. Here again, we consider only the most dominant balance, meaning that none of the coefficients α of the prefactor term in (7.2) is arbitrary. The NTT acts then on these solutions in the following way:

$$p_i \rightarrow \frac{p_i}{1+c} \tag{8.7}$$

$$r \rightarrow \begin{cases} -1 & \text{for one resonance } r = -1 \\ \frac{r}{1+c} & \text{otherwise} \end{cases} \tag{8.8}$$

with $c = -\sum_{i=1}^{n} p_i \beta_i$.

Let us lay stress here on the above transformation rules of the dominant behaviour and the resonance set. We see that if the solution contains algebraic singularities and can be expanded in Puiseux series, then by a suitable choice of an NTT, it is mapped onto a Laurent expansion. On the other hand, the drawback of this mapping is that every Puiseux expansion should be simultaneously mapped onto a Laurent series. This is due to the fact that the NTT is a local transformation including the global knowledge of the solution $x = x(t)$ near every singularity. It can be shown that for the integrable weak-Painlevé systems, the NTT provides a procedure for finding the new coordinates in which the system recovers its Painlevé property. Nevertheless, a general connection between Painlevé and weak-Painlevé systems via the NTTs cannot be found due to the existence of nonintegrable weak-Painlevé systems which are not integrable.

Let us now consider the following system:

$$H = \frac{1}{2}(p_x^2 + p_y^2) - \frac{1}{\gamma}(x^2y^2)^\gamma \qquad (8.9)$$

with $\gamma \geq 0$

It can be shown that for all $\gamma \in N$, this system is exactly nonintegrable in the sense of Ziglin, that is there is no other complex analytic constant of motion. Moreover, using symbolic dynamics, this system was proved to exhibit chaotic motion rendering the search for integrable domains of the control parameter γ hopeless. However, this system is of weak-Painlevé type for an infinite set of values γ, the first few ones being:

$$\gamma \in \{23/12, \ 77/40, \ 109/56, \ 233/112, \ 43/20, \ 53/24, ...\}$$

This counterexample does not allow for a rigorous treatment of algebraic singularities and integrability except in the case where an NTT can be defined for a weak-Painlevé system. In the nonintegrable domain, the role of algebraic singularities and the interaction between singularities is still not understood since it has been shown that the interaction between two square-root branch type singularities can give rise to chaotic behaviour according to numerical experiments[16].

We hope to have shown that the quasi-monomial approach although based on a quite restricted set of transformations, the QM and NTT transformations, offers both a deep insight into the behaviour of nonlinear dynamical systems and a powerful constructive computation tool. Indeed, in this framework, the behaviours of a wide class of systems may be exactly reduced to those of the Lotka-Volterra systems. Relations are found between disparate properties such as exact decoupling and normal forms, the existence of the latter and the locations of complex singularities. In the near future we shall explore these paths further as well as other directions suggested by the present work: the connection between bifurcations and equilibrium phase transitions, the frontier separating integrable and nonintegrable systems. Moreover, the fact that, in analogy with the linear differential systems, the nonlinear systems are here characterized by objects as simple as matrices allows for a thorough translation of the related algorithms into efficient computer algebra programs. These are constituting the software NODES whose development in Brussels follows closely the progress of the theoretical work.

References

[1] Br'uno, *Local Methods in the Theory of Differential Equations*, Springer Verlag, 1989.

[2] Peschel M., Mende, W., *Leben wir in einer Volterra-Welt.* ,Mathematical Research 14, Akademie Verlag, Berlin, 1983

[3] L. Brenig, *Phys. Let. A133 (1988)*, pp. 378-382

[4] Brenig L. and Goriely A., *Phys. Rev. A40 (1989)*, pp. 4119-4122

[5] Goriely A. and Brenig, L., *Phys. Let. A145 (1990)*, p. 245

[6] Arnold, V.I., and Ilyaschenko Y.S., *Ordinary Differential Equations* in *Dynamical Systems I, D.V. Anosov and V.I. Arnold Eds*, Springer Verlag, 1988

[7] Rocha Filho, T. M. and Brenig, L.,*Submitted for Publication, 1992*

[8] May R. M., *Stability and Complexity in Model Ecosystems*, Princeton University Press, Princeton, N.J.,1973

[9] Arneodo A., Coullet P. and Tresser C., *Phys. Let. 79A (1981)*, pp. 259-263

[10] Parisi G., *Statistical Field Theory*, Addison-Wesley, New York, 1988

[11] Painlevé P., *Acta Math 25 (1900) p.1*

[12] Kowaleskaya, S., *Acta Math. 12 (1989) p.177*

[13] McLeod and Olver P. J., *SIAM J. Math. Anal. 14 (1983)*, pp.488-506

[14] Adler M. and van Moerbecke P., *Invent. Math. 97 (1989) p.3*

[15] Goriely A., *J. Math. Phys. 33 (1992)*, pp.2728-2742

[16] Bountis T, *Int. J. of Bif. and Chaos 2 (1992)*, p.217

Normal Forms and Stokes Multipliers of Nonlinear Meromorphic Differential Equations

Yasutaka Sibuya *

Contents

1. Preliminaries 240

2. Main problem 242

3. A formal solution by means of a formal normal form 245

4. A normal form of a linear system 248

5. Stokes multipliers 252

6. Proof of Theorem 2.1 258

7. Acknowledgements 258

8. References 259

*School of Mathematics, University of Minnesota, Minneapolis, Minnesota 55455, U.S.A.

Introduction

Multisummability of formal solutions of meromorphic differential equations was proved by J.-P.Ramis [9], J.Martinet and J.-P.Ramis [8], B.Malgrange and J.-P.Ramis [7], B.L.J.Braaksma [2,3] and W.Balser, B.L.J.Braaksma, J.-P.Ramis and Y.Sibuya [1]. In particular, B.L.J.Braaksma [3] treated nonlinear cases by means of a method based on J.Ecalle's theory of acceleration (cf. J.Ecalle [4] and J.Martinet and J.-P.Ramis [8]). In this paper, we shall outline another proof based on the cohomological definition of multisummability (cf. B.Malgrange and J.-P.Ramis [7] and W.Balser, B.L.J.Braaksma, J.-P.Ramis and Y.Sibuya [1]). The main problem is explained in §2 (cf. Theorem 2.1). In this paper, we shall outline a proof of Theorem 2.1 only, since multisummability of formal power series solutions can be derived from Theorem 2.1 in a manner similar to the proof of Theorem 4.1 based on Lemma 7.1 in paper [1]. In our outline, we shall show mostly the formal part which is the key idea. An analytic justification of the formal part utilizes methods due to M.Hukuhara [5], M.Iwano [6], and J.-P.Ramis and Y.Sibuya [10]. We shall publish another paper (jointly with J.-P.Ramis) in which the entire analysis will be explained in detail.

1.   Preliminaries

Throughout this paper we shall use the following notations:

1) $\lambda(x)$ is an $n \times n$ diagonal matrix:

$$\lambda(x) = \begin{bmatrix} \lambda_1(x) & 0 & 0 & \cdots & 0 & 0 \\ 0 & \lambda_2(x) & 0 & \cdots & \cdots & 0 \\ 0 & 0 & \lambda_3(x) & \cdots & 0 & 0 \\ \cdots & \cdots & \cdots & \cdots & \cdots & \cdots \\ \cdots & \cdots & \cdots & \cdots & \cdots & \cdots \\ 0 & 0 & 0 & \cdots & 0 & \lambda_n(x) \end{bmatrix}, \qquad (1.1)$$

where either $\lambda_j(x) = 0$ identically or

$$\lambda_j(x) = \sum_{\ell=1}^{N_j} \lambda_{j,\ell}\, x^{-\nu_\ell} \,; \qquad (1.2)$$

here

$$\begin{cases} 0 < \nu_1 < \nu_2 < \cdots < \nu_{\tilde{N}} \,, \\ 1 \le N_j \le \tilde{N} \,, \\ \lambda_{j,\ell} \in C \,, \\ \lambda_{j,N_j} \ne 0 \,. \end{cases} \qquad (1.3)$$

2)

$$A_0 = \begin{bmatrix} \mu_1 & \delta_1 & 0 & \cdots & 0 & 0 \\ 0 & \mu_2 & \delta_2 & \cdots & 0 & 0 \\ \cdots & \cdots & \cdots & \cdots & \cdots & \cdots \\ 0 & 0 & 0 & \cdots & \mu_{n-1} & \delta_{n-1} \\ 0 & 0 & 0 & \cdots & 0 & \mu_n \end{bmatrix}, \qquad (1.4)$$

where

(i) the μ_j are complex numbers such that the differences $\mu_j - \mu_{j'}$ are not equal to nonzero integers if $\lambda_j(x) = \lambda_{j'}(x)$;

(ii) the δ_j are complex numbers such that $\lambda_j(x) + \mu_j = \lambda_{j+1}(x) + \mu_{j+1}$ if $\delta_j \neq 0$;

(iii) $\Re(\mu_j) < 0$ $(j = 1, 2, \cdots, n)$.

3) The quantity μ is a positive number such that $\mu + \Re(\mu_j - \mu_h) > 0$ $(j, h = 1, 2, \cdots, n)$.

4) We denote by $\mathcal{R}$ the set of all j such that $\lambda_j(x)$ is not identically equal to zero; i.e.

$$\mathcal{R} = \{ j ; \lambda_j(x) \neq 0 \} . \qquad (1.5)$$

5) We set

$$\Lambda_j(x) = \begin{cases} 0 & \text{if } j \notin \mathcal{R} , \\ \displaystyle\sum_{\ell=1}^{N_j} \frac{-\lambda_{j,\ell}}{\nu_\ell} x^{-\nu_\ell} & \text{if } j \in \mathcal{R} . \end{cases} \qquad (1.6)$$

6) We also set

$$\tau_j = \nu_{N_j} \quad \text{for } j \in \mathcal{R} . \qquad (1.7)$$

7) Let

$$0 < k_p < k_{p-1} < \cdots < k_2 < k_1 < +\infty \qquad (1.8)$$

be all of the distinct real numbers in the set $\{ \tau_j ; j \in \mathcal{R} \}$; i.e.

$$\{ k_1, \cdots, k_p \} = \{ \tau_j ; j \in \mathcal{R} \}.$$

8) We fix an integer q such that $2 \leq q \leq p$ and set

$$k = k_q, \qquad k' = k_{q-1}. \qquad (1.9)$$

9) We also set

$$\mathcal{R}_k = \{ j \in \mathcal{R} ; \tau_j = k \} . \qquad (1.10)$$

Remark: Throughout this paper, all sectorial domains are considered on the Riemann surface of log x. Also, for an m-vector $\vec{y} = \begin{bmatrix} y_1 \\ \cdots \\ y_m \end{bmatrix}$, we define a norm $|\vec{y}|$ by

$$|\vec{y}| = \max_{1 \le j \le m} |y_j| . \tag{1.11}$$

Furthermore, for every $\wp = (p_1, \cdots, p_m)$ where the p_ℓ are nonnegative integers, we set

$$|\wp| = p_1 + p_2 + \cdots + p_m, \qquad \vec{y}^\wp = y_1^{p_1} y_2^{p_2} \cdots y_m^{p_m} . \tag{1.12}$$

2.   Main problem

We consider a differential equation:

$$x \frac{d\vec{y}}{dx} = \vec{G}_0(x) + [\lambda(x) + A_0] \vec{y} + x^\mu \vec{G}(x, \vec{y}) , \tag{2.1}$$

where

(I) the n-vector $\vec{G}_0(x)$ is holomorphic in an open sector
$\mathcal{D}(a, b, r_0) = \{ x ; a < \arg x < b, 0 < |x| < r_0 \}$;

(II) for every closed subsector $\mathcal{D}[\alpha, \beta, r] = \{ x ; \alpha \le \arg x \le \beta, 0 < |x| \le r \}$ of $\mathcal{D}(a, b, r_0)$, there exists a positive number $\rho(\alpha, \beta, r)$ such that the power series

$$\vec{G}(x, \vec{y}) = \sum_{|\wp| \ge 1} \vec{y}^\wp \vec{G}_\wp(x)$$

is uniformly convergent for

$$x \in \mathcal{D}[\alpha, \beta, r], \qquad |\vec{y}| \le \rho(\alpha, \beta, r) , \tag{2.2}$$

where the coefficients $\vec{G}_\wp$ are holomorphic and bounded in $\mathcal{D}(a, b, r_0)$.

We assume the following conditions:

(1) there exists a direction $\arg x = d$ such that

$$a < d - \frac{\pi}{2k} < d + \frac{\pi}{2k} < b ; \tag{2.3}$$

(2) for each $j \in \mathcal{R}_k$, there exists a direction $\arg x = d_j$ such that

$$d - \frac{\pi}{2k} < d_j < d + \frac{\pi}{2k} \tag{2.4}$$

and that $\Re[\Lambda_j(x)]$ changes its sign across the direction $\arg x = d_j$ (cf. figure 1). This means that the direction $\arg x = d$ is not singular on the level k.

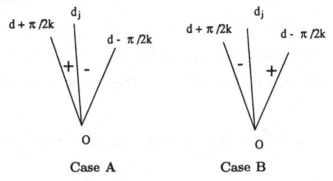

Case A   **Case B**

Figure 1

the sign of $\Re[\Lambda_j(x)]$.

We consider the following situation: for any positive number r, set

$$\mathcal{W}_0(r) = \{ x \in C ; \mid \arg x - d \mid \leq \frac{\pi}{2k}, \; 0 < \mid x \mid < r \} \qquad (2.5)$$

and let

$$\mathcal{U}_\nu(r) = \mathcal{D}(\alpha_\nu, \beta_\nu, r) \qquad (\nu = 1, 2, \cdots, N)$$

be a covering of $\mathcal{W}_0(r)$, i.e.

$$\mathcal{W}_0(r) \subset \bigcup_{1 \leq \nu \leq N} \mathcal{U}_\nu(r) \subset \mathcal{D}\left(d - \frac{\pi}{2k} - \epsilon_0, d + \frac{\pi}{2k} + \epsilon_0, r\right), \qquad (2.6)$$

where ϵ_0 is a sufficiently small positive number. Let us assume that the covering $\{ \mathcal{U}_\nu(r) ; \; \nu = 1, 2, \cdots, N \}$ satisfies the following conditions:

(i) $0 < \beta_\nu - \alpha_\nu < \frac{\pi}{k_1} \qquad (\nu = 1, 2, \cdots, N)$;

(ii) $a < \alpha_1 < \alpha_2 < \cdots < \alpha_N$ and $\beta_1 < \beta_2 < \cdots < \beta_N < b$;

(iii)
$$\mathcal{U}_\nu(r) \cap \mathcal{U}_{\nu'}(r) \begin{cases} \neq \emptyset & \text{if } \mid \nu - \nu' \mid \leq 1, \\ = \emptyset & \text{if } \mid \nu - \nu' \mid \geq 2; \end{cases}$$

(iv) there exists a positive number r_1 and N functions $\vec{f}_1(x), \cdots, \vec{f}_N(x)$ such that

(a) for each ν, the function $\vec{f}_\nu$ is holomorphic in $\mathcal{U}_\nu(r_1)$;

(b) for each ν, we have $\lim \vec{f}_\nu(x) = \vec{0}$ as $x \to 0$ in $\mathcal{U}_\nu(r_1)$;

(c) for each $\nu \geq 2$, we have

$$\left| \vec{f}_\nu(x) - \vec{f}_{\nu-1}(x) \right| \leq K\, e^{-\epsilon_1 |\,x\,|^{-k}} \quad \text{in} \quad \mathcal{U}_\nu(r_1) \cap \mathcal{U}_{\nu-1}(r_1)$$
(2.7)

for some positive numbers K and ϵ_1 ;

(d) for each ν, $\vec{f}_\nu$ is a solution of differential equation (2.1) in $\mathcal{U}_\nu(r_1)$, i.e.

$$x\,\frac{d\vec{f}_\nu(x)}{dx} = \vec{G}_0(x) + [\,\lambda(x) + A_0\,]\,\vec{f}_\nu(x) + x^\mu\,\vec{G}(x, \vec{f}_\nu(x)) \quad (2.8)$$

$$\text{in} \quad \mathcal{U}_\nu(r_1).$$

Remark: we can choose the α_ν and the β_ν so that, if $\Re\,[\Lambda_j(x)] \leq 0$ on

$$\mathcal{W}_\nu(r) = \{\, x;\ \alpha_\nu \leq \arg x \leq \beta_\nu,\ 0 < |\,x\,| \leq r \,\} \quad (2.9)$$

for sufficiently small $r > 0$, then $\Re\,[\Lambda_j(x, \epsilon; \nu)] < 0$ on $\mathcal{W}_\nu(r)$ for sufficiently small $r > 0$. In particular, we can choose the α_ν and the β_ν so that

$$\alpha_\nu \neq d_j \quad \text{and} \quad \beta_\nu \neq d_j \quad \text{for} \quad j \in \mathcal{R}_k,\ \nu = 1, 2, \cdots, N.$$

Also we can assume that the directions $\arg x = d_j$ are not in $\mathcal{U}_\nu(r_1) \cap \mathcal{U}_{\nu-1}(r_1)$ for any ν (≥ 2).

Now, we can state the main result of this paper.

Theorem 2.1. *We can modify the N functions $f_1, \cdots, f_N$ by some quantities of $O\left(e^{-\epsilon|x|^{-k}}\right)$, where ϵ is some positive constant, so that these modified functions also satisfy conditions (a),(b) and (d) of (iv) given above and that moreover they satisfy the following condition (c'):*

(c') *for each $\nu \geq 2$, we have*

$$\left| \vec{f}_\nu(x) - \vec{f}_{\nu-1}(x) \right| \leq K'\, e^{-\epsilon_2 |\,x\,|^{-k'}} \quad \text{in} \quad \mathcal{U}_\nu(r_2) \cap \mathcal{U}_{\nu-1}(r_2) \quad (2.10)$$

for some positive numbers K', r_2 and ϵ_2.

3.   A formal solution by means of a formal normal form

We consider a differential equation,

$$x\frac{d\vec{u}}{dx} = [\,\lambda(x) + A_0\,]\vec{u} + x^{\mu}\vec{F}(x,\vec{u}), \qquad \vec{u} = \begin{bmatrix} u_1 \\ u_2 \\ \cdots \\ u_n \end{bmatrix}, \qquad (3.1)$$

where $\vec{F}$ is a power series in $u_1, \cdots, u_n$:

$$\vec{F}(x,\vec{u}) = A(x)\vec{u} + \sum_{|p|\geq 2} \vec{u}^p\,\vec{F}_p(x)\,, \qquad (3.2)$$

satisfying the following conditions:

(i) $A(x)$ is an $n \times n$ matrix whose entries are holomorphic and bounded in a sectorial domain,
$$\mathcal{D}(\alpha,\beta,r) = \{\,x\,;\,\alpha < \arg x < \beta,\, 0 < |x| < r\,\}\,, \qquad (3.3)$$

(ii) the $\vec{F}_p(x)$ are n-vectors whose entries are holomorphic and bounded in $\mathcal{D}(\alpha,\beta,r)\,,$

(iii) the power series $\vec{F}$ is uniformly convergent for
$$x \in \mathcal{D}(\alpha,\beta,r), \qquad |\,\vec{u}\,| < 2\rho\,. \qquad (3.4)$$

Remark: There exists a nonnegative number L such that
$$\left|\,\vec{F}(x,\vec{u}) - \vec{F}(x,\vec{u}')\,\right| \leq L|\,\vec{u} - \vec{u}'\,| \qquad (3.5)$$

for
$$x \in \mathcal{D}(\alpha,\beta,r), \qquad |\,\vec{u}\,| < \rho\,, \qquad |\,\vec{u}'\,| < \rho\,. \qquad (3.6)$$

In particular,
$$\left|\,\vec{F}(x,\vec{u})\,\right| \leq L|\,\vec{u}\,| \qquad (3.7)$$

for
$$x \in \mathcal{D}(\alpha,\beta,r), \qquad |\,\vec{u}\,| < \rho\,. \qquad (3.8)$$

Let us also assume the following conditions:

(iv) $\beta - \alpha < \dfrac{\pi}{k_1}\,;$

(v) *if, for some $j \in \mathcal{R}$, $\Re[\Lambda_j(x)] \leq 0$ in $\mathcal{D}(\alpha, \beta, r)$, then $\Re[\Lambda_j(x)] \leq -\delta$ in $\mathcal{D}(\alpha, \beta, r)$ for some positive number δ ;*

(vi)

$$\mathcal{J}_k = \{ j \in \mathcal{R}_k \, ; \, \Re[\Lambda_j(x)] < 0 \text{ in } \mathcal{D}(\alpha, \beta, r) \} = \{ 1, \cdots, n_0 \} . \quad (3.9)$$

Let us set:

$[\lambda(x) + A_0]_{n_0}$

$$= \begin{bmatrix} \lambda_1(x) + \mu_1 & \delta_1 & 0 & \cdots & 0 & 0 \\ 0 & \lambda_2(x) + \mu_2 & \delta_2 & \cdots & 0 & 0 \\ \cdots & \cdots & \cdots & \cdots & \cdots & \cdots \\ 0 & 0 & 0 & \cdots & \lambda_{n_0-1}(x) + \mu_{n_0-1} & \delta_{n_0-1} \\ 0 & 0 & 0 & \cdots & 0 & \lambda_{n_0}(x) + \mu_{n_0} \end{bmatrix}$$

$$(3.10)$$

and

$$\vec{w} = \begin{bmatrix} w_1 \\ w_2 \\ \cdots \\ w_{n_0} \end{bmatrix} .$$

Lemma 3.1. *For every $\wp = (p_1, \cdots, p_{n_0})$ where the p_ℓ are nonnegative integers such that $|\wp| = p_1 + \cdots + p_{n_0} \geq 2$, we can find an n-vector $\vec{P}_\wp(x)$ and an n_0-vector $\vec{\alpha}_\wp$:*

$$\vec{P}_\wp(x) = \begin{bmatrix} P_{\wp,1}(x) \\ \cdots \\ P_{\wp,n}(x) \end{bmatrix}, \qquad \vec{\alpha}_\wp = \begin{bmatrix} \alpha_{\wp,1}(x) \\ \cdots \\ \alpha_{\wp,n_0}(x) \end{bmatrix},$$

together with an $n \times n_0$ matrix $P_0(x)$ in such a way that

(i) *the matrix $P_0(x)$ is holomorphiuc in $\mathcal{D}(\alpha, \beta, r)$ and*

$$|x|^{-\mu'} |P_0(x) - C| \qquad \left(C = \begin{bmatrix} I_{n_0} \\ O \end{bmatrix} \right)$$

is bounded in $\mathcal{D}(\alpha, \beta, r)$ for some positive number μ' such that

$$\mu' + \Re[\mu_j - \mu_h] > 0 \qquad (\text{for } j, h = 1, \cdots, n),$$

where I_{n_0} is the $n_0 \times n_0$ identity matrix and O is the $(n - n_0) \times n_0$ zero matrix;

(ii) the $\vec{P}_p(x)$ and $\vec{\alpha}_p(x)$ are holomorphic and bounded in $\mathcal{D}(\alpha,\beta,r)$;

(iii) $P_{p,j}(x) = 0$ if $\lambda_j(x) = \sum_{1\le\ell\le n_0} p_\ell \lambda_\ell(x)$;

(iv) $\alpha_{p,j}(x) = 0$ if $\lambda_j(x) \ne \sum_{1\le\ell\le n_0} p_\ell \lambda_\ell(x)$;

(v) the formal power series

$$\vec{P}(x,\vec{w}) \;=\; P_0(x)\vec{w} \;+\; x^\mu \sum_{|p|\ge 2} \vec{w}^p \vec{P}_p(x) \tag{3.11}$$

is a formal solution of differential equation (3.1) if

$$x\frac{d\vec{w}}{dx} \;=\; [\lambda(x) + A_0]_{n_0}\vec{w} \;+\; x^\mu \sum_{|p|\ge 2} \vec{w}^p \vec{\alpha}_p(x) . \tag{3.12}$$

In order to find some meaning of formal solution (3.11) of Lemma 3.1, let us first look at differential equation (3.12). Since condition (iv) of Lemma 3.1 is satisfied, differential equation (3.12) becomes

$$x\frac{d\vec{v}}{dx} \;=\; [\,A_0\,]_{n_0}\vec{v} \;+\; x^\mu \sum_{|p|\ge 2} \vec{v}^p \vec{\alpha}_p(x) , \tag{3.13}$$

if we set

$$w_j \;=\; e^{\Lambda_j(x)}\,v_j \qquad (j = 1,\cdots,n_0) , \tag{3.14}$$

where

$$[\,A_0\,]_{n_0} = \begin{bmatrix} \mu_1 & \delta_1 & 0 & \cdots & 0 & 0 \\ 0 & \mu_2 & \delta_2 & \cdots & 0 & 0 \\ \cdots & \cdots & \cdots & \cdots & \cdots & \cdots \\ 0 & 0 & 0 & \cdots & \mu_{n_0-1} & \delta_{n_0-1} \\ 0 & 0 & 0 & \cdots & 0 & \mu_{n_0} \end{bmatrix}. \tag{3.15}$$

Let $\vec{c} = \begin{bmatrix} c_1 \\ c_2 \\ \cdots \\ c_{n_0} \end{bmatrix}$ be an arbitrary constant n_0-vector. If we arrange the $\Lambda_j(x)$ in such a way that

$$\Re\,[\Lambda_1(x)] \le \Re\,[\Lambda_2(x)] \le \cdots \le \Re\,[\Lambda_{n_0}(x)]$$

in a direction $\arg x = \theta$ in $\mathcal{D}(\alpha,\beta,r)$, a general solution of (3.13) is given by

$$v_j \;=\; \psi_j(x;\vec{c}) \;=\; c_j x^{\mu_j} + \tilde{\psi}_j(x;c_{j+1},...,c_{n_0}) \qquad (j = 1,\cdots,n_0) , \tag{3.16}$$

where the $\tilde{\psi}_j$ are holomorphic in $\mathcal{D}(\alpha,\beta,r)$ and

$$\tilde{\psi}_j \;=\; \sum_{|p|\ge 1} \vec{c}^{\,p}\, \tilde{\psi}_{jp}(x) , \tag{3.17}$$

where

$$\tilde{\psi}_{jp}(x) = 0 \qquad \text{if} \quad \lambda_j \neq \sum_{\ell=1}^{n_0} p_\ell \lambda_\ell \tag{3.18}$$

and

$$| \tilde{\psi}_{jp}(x) | \leq \gamma_0 | \, x \, |^{\mu_0} \tag{3.19}$$

in $\mathcal{D}(\alpha, \beta, r)$ for some positive number γ_0 and some real number μ_0. In fact, the ψ_j are polynomials in $c_1, \cdots, c_{n_0}$.

Putting (3.14) and (3.16) together, we can find a general solution of (3.12), i.e.

$$\vec{w} = \vec{\phi}(x; \vec{c}) = \begin{bmatrix} \phi_1(x; \vec{c}) \\ \phi_2(x; \vec{c}) \\ \cdots \\ \phi_{n_0}(x; \vec{c}) \end{bmatrix}, \qquad \phi_j(x; \vec{c}) = e^{\Lambda_j(x)} \psi_j(x; \vec{c}) \quad (j = 1, \cdots, n_0) \ .$$

$$\tag{3.20}$$

By utilizing (3.18), we can write $\vec{\phi}(x; \vec{c})$ in the following form:

$$\vec{\phi}(x; \vec{c}) = x^{[A_0]_{n_0}} e^{[\Lambda(x)]_{n_0}} \vec{c} + \sum_{| \, p \, | \geq 2} \left(e^{[\Lambda(x)]_{n_0}} \vec{c} \right)^{p} \vec{\phi}_p(x) \ , \tag{3.21}$$

where

$$[\Lambda(x)]_{n_0} = diag(\Lambda_1(x), \cdots, \Lambda_{n_0}(x))$$

and

$$| \vec{\phi}_p(x) | \leq \gamma_0 | \, x \, |^{\mu_0} \tag{3.22}$$

for some positive number γ_0 and some real number μ_0 in $\mathcal{D}(\alpha, \beta, r)$.

Remark: For a proof of Lemma 3.1, see M.Hukuhara[5] and M.Iwano[6].

4.   A normal form of a linear system

Let us assume that

$\{1, \cdots, n_1\} \in \mathcal{R}_k$ and $\{n_0 + 1, \cdots, n_0 + n_2\} \in \mathcal{R}_k$ and that

$$\Re \, [\Lambda_j(x)] < 0 \qquad \text{in} \quad \mathcal{D}(\alpha, \beta, r) \tag{4.1}$$

for $j = 1, \cdots, n_1$ and $j = n_0 + 1, \cdots, n_0 + n_2$. Let

$$\vec{\phi}_1(x; \vec{c}_1) = \begin{bmatrix} \phi_1(x; \vec{c}_1) \\ \phi_2(x; \vec{c}_1) \\ \cdots \\ \phi_{n_1}(x; \vec{c}_1) \end{bmatrix}, \qquad \vec{\phi}_2(x; \vec{c}_2) = \begin{bmatrix} \tilde{\phi}_{n_0+1}(x; \vec{c}_2) \\ \tilde{\phi}_{n_0+2}(x; \vec{c}_2) \\ \cdots \\ \tilde{\phi}_{n_0+n_2}(x; \vec{c}_2) \end{bmatrix}$$

be general solutions of the following two differential equations:

$$x\frac{d\vec{w}_1}{dx} = [\,\lambda(x) + A_0\,]_1\,\vec{w}_1 \;+\; x^\mu \sum_{|\wp|\geq 2} \vec{w}_1^\wp\,\vec{\alpha}_{1,\wp}(x)$$

and

$$x\frac{d\vec{w}_2}{dx} = [\,\lambda(x) + A_0\,]_2\,\vec{w}_2 \;+\; x^\mu \sum_{|\wp|\geq 2} \vec{w}_2^\wp\,\vec{\alpha}_{2,\wp}(x)\;,$$

respectively where

$$[\,\lambda(x) \;+\; A_0\,]_1$$

$$= \begin{bmatrix} \lambda_1(x)+\mu_1 & \delta_1 & 0 & \cdots & 0 \\ 0 & \lambda_2(x)+\mu_2 & \delta_2 & \cdots & 0 \\ \cdots & \cdots & \cdots & \cdots & \cdots \\ 0 & 0 & 0 & \cdots & \delta_{n_1-1} \\ 0 & 0 & 0 & \cdots & \lambda_{n_1}(x)+\mu_{n_1} \end{bmatrix}$$

and

$$[\,\lambda(x) \;+\; A_0\,]_2$$

$$= \begin{bmatrix} \lambda_{n_0+1}(x)+\mu_{n_0+1} & \delta_{n_0+1} & 0 & \cdots & & 0 & \cdots \\ 0 & & & & & & \\ \cdots & \cdots & \cdots & \cdots & & \cdots & \\ 0 & 0 & 0 & \cdots & \delta_{n_0+n_2-1} & & \\ 0 & 0 & 0 & \cdots & \lambda_{n_0+n_2}(x)+\mu_{n_0+n_2} & & \end{bmatrix}\;.$$

Set

$$\vec{\alpha}_{1,\wp}(x) = \begin{bmatrix} \alpha_{1,\wp,1}(x) \\ \alpha_{1,\wp,2}(x) \\ \cdots \\ \alpha_{1,\wp,n_1}(x) \end{bmatrix}, \qquad \vec{\alpha}_{2,\wp}(x) = \begin{bmatrix} \alpha_{2,\wp,n_0+1}(x) \\ \alpha_{2,\wp,n_0+2}(x) \\ \cdots \\ \alpha_{2,\wp,n_0+n_2}(x) \end{bmatrix}\;.$$

We assume that

$$\begin{cases} \alpha_{1,\wp_1,j}(x) \;=\; 0 & \text{if } \lambda_j \neq \displaystyle\sum_{1\leq\ell\leq n_1} p_{1,\ell}\lambda_\ell\;, \\[4mm] \alpha_{2,\wp_2,n_0+j}(x) = 0 & \text{if } \lambda_{n_0+j} \neq \displaystyle\sum_{1\leq\ell\leq n_2} p_{2,n_0+\ell}\lambda_{n_0+\ell}\;, \end{cases} \tag{4.2}$$

where $\wp_1 = (p_{1,1}, \cdots, p_{1,n_1})$ and $\wp_2 = (p_{2,n_0+1}, \cdots, p_{2,n_0+n_2})$.

Lemma 4.1. *Let an $n \times n$ matrix*

$$A(x, \vec{\phi}_1, \vec{\phi}_2) = \sum_{|\wp_1|+|\wp_2|\geq 0} \vec{\phi}_1^{\wp_1} \vec{\phi}_2^{\wp_2} A_{\wp_1\wp_2}(x) \qquad (4.3)$$

be given as a power series in $\vec{\phi}_1$ and $\vec{\phi}_2$, where the $A_{\wp_1\wp_2}(x)$ are holomorphic and bounded in $\mathcal{D}(\alpha, \beta, r)$. Then for every $\wp_1 = (p_{1,1}, \cdots, p_{1,n_1})$ and $\wp_2 = (p_{2,n_0+1}, \cdots, p_{2,n_0+n_2})$, where $|\wp_1| + |\wp_2| \geq 1$, we can find $n \times n$ matrices $\Phi_{\wp_1\wp_2}(x)$ and $B_{\wp_1\wp_2}(x)$ together with another $n \times n$ matrix $\Phi_0(x)$ in such a way that

(i) $\Phi_{\wp_1\wp_2}(x)$, $B_{\wp_1\wp_2}(x)$ *and $\Phi_0(x)$ are holomorphic and bounded in $\mathcal{D}(\alpha,\beta,r)$;*

(ii) *if we denote by $\Phi_{\wp_1\wp_2,jj'}(x)$ and $B_{\wp_1\wp_2,jj'}(x)$ the (j,j')-th entries of $\Phi_{\wp_1\wp_2}$ and $B_{\wp_1\wp_2}$ respectively, then*

$$\begin{cases} \Phi_{\wp_1\wp_2,jj'}(x) = 0 \quad if \quad \lambda_j - \lambda_{j'} = \sum_{1\leq\ell\leq n_1} p_{1,\ell}\lambda_\ell + \sum_{1\leq\ell\leq n_2} p_{2,n_0+\ell}\lambda_{n_0+\ell} , \\[4mm] B_{\wp_1\wp_2,jj'}(x) = 0 \quad if \quad \lambda_j - \lambda_{j'} \neq \sum_{1\leq\ell\leq n_1} p_{1,\ell}\lambda_\ell + \sum_{1\leq\ell\leq n_2} p_{2,n_0+\ell}\lambda_{n_0+\ell} ; \end{cases}$$

$$(4.4)$$

(iii) *there is a positive number μ' such that $\mu' + \Re[\mu_j - \mu_{j'}] > 0$ for $j, j' = 1, \cdots, n$ and that, if we put*

$$\begin{cases} \Phi(x, \vec{\phi}_1, \vec{\phi}_2) = I + x^{\mu'}\Phi_0 + x^\mu \sum_{|\wp_1|+|\wp_2|\geq 1} \vec{\phi}_1^{\wp_1}\vec{\phi}_2^{\wp_2}\Phi_{\wp_1\wp_2}, \\[4mm] B(x, \vec{\phi}_1, \vec{\phi}_2) = \sum_{|\wp_1|+|\wp_2|\geq 1} \vec{\phi}_1^{\wp_1}\vec{\phi}_2^{\wp_2} B_{\wp_1\wp_2}, \end{cases}$$

$$(4.5)$$

we have

$$x\frac{d\Phi}{dx} = [\,\lambda(x) + A_0 + x^\mu B]\,\Phi - \Phi[\,\lambda(x) + A_0 + x^\mu A] \qquad (4.6)$$

as formal power series in $\vec{\phi}_1$ and $\vec{\phi}_2$.

Observation 4.2: We can construct a fundamental matrix $\Psi(x, \vec{c}_1, \vec{c}_2)$ of the system

$$x\frac{d\vec{V}}{dx} = -\vec{V}[\,\lambda(x) + A_0 + x^\mu B]\,, \qquad (4.7)$$

where $\vec{V}$ is an n-dimensional row vector, in such a way that

$$e^{\Lambda(x)}\Psi(x,\vec{c}_1,\vec{c}_2) = \sum_{|p_1|+|p_2|\geq 0} \left(e^{[\Lambda]_1}\vec{c}_1\right)^{p_1} \left(e^{[\Lambda]_2}\vec{c}_2\right)^{p_2} \Psi_{p_1 p_2}(x)$$

where

$$\begin{cases} \Lambda(x) = diag\,(\Lambda_1(x),\cdots,\Lambda_n(x)) \ , \\ [\Lambda(x)]_1 = diag\,(\Lambda_1(x),\cdots,\Lambda_{n_1}(x)) \ , \\ [\Lambda(x)]_2 = diag\,(\Lambda_{n_0+1}(x),\cdots,\Lambda_{n_0+n_2}(x)) \ , \end{cases} \tag{4.8}$$

and the $\Psi_{p_1 p_2}(x)$ are holomorphic in $\mathcal{D}(\alpha,\beta,r)$; furthermore

$$|\,\Psi_{p_1 p_2}(x)\,| \leq \gamma_{p_1 p_2}\,|\,x\,|^{\mu_{p_1 p_2}} \tag{4.9}$$

in $\mathcal{D}(\alpha,\beta,r)$ for some positive numbers $\gamma_{p_1 p_2}$ and some real numbers $\mu_{p_1 p_2}$.

Observation 4.3: More precisely speaking, if we set

$$\vec{V} = \vec{W}x^{-A_0}e^{-\Lambda(x)} \ , \tag{4.10}$$

differential equation (4.7) changes to

$$x\frac{d\vec{W}}{dx} = -\vec{W}\,x^{\mu}x^{-A_0}\tilde{B}x^{A_0} \ , \tag{4.11}$$

where

$$\tilde{B} = e^{-\Lambda(x)}Be^{\Lambda(x)} \ . \tag{4.12}$$

Since the matrix B satisfies condition (4.4), we have

$$|\,\tilde{B}\,| \leq \gamma_0\,|\,x\,|^{\mu_0} \tag{4.13}$$

in $\mathcal{D}(\alpha,\beta,r)$ for some positive number γ_0 and some real number μ_0 . Furthermore if we assume that

$$\Re[\Lambda_1(x)] \leq \Re[\Lambda_2(x)] \leq \cdots \leq \Re[\Lambda_n(x)] \ , \tag{4.14}$$

and if we denote by $\tilde{B}_{jj'}$ the (j,j')-th entry of $\tilde{B}$, we have

$$\tilde{B}_{jj'} = 0 \quad \text{if} \quad \Re[\Lambda_j(x)] \geq \Re[\Lambda_{j'}(x)] \ . \tag{4.15}$$

The matrix $\Psi(x,\vec{c}_1,\vec{c}_2)$ has the following form:

$$\Psi(x,\vec{c}_1,\vec{c}_2) = \left[I + \int^x \xi^{\mu}\xi^{-A_0}\tilde{B}(\xi,\vec{c}_1,\vec{c}_2)\xi^{A_0}d\xi + \cdots\right]x^{-A_0}e^{-\Lambda(x)} \ , \tag{4.16}$$

where $\cdots$ denotes terms of $O(|\vec{c}_1|^2 + |\vec{c}_2|^2)$.

Remark: Lemma 4.1 can be proved in a manner similar to the proof of Lemma 3.1.

5.   Stokes multipliers

We shall utilize the same notations as in §2,

and consider the following situation: let

$$\mathcal{U}_\nu(r) \; = \; \mathcal{D}(\alpha_\nu, \beta_\nu, r) \qquad (\nu = 1, 2)$$

be two subsectors of $\mathcal{D}(a, b, r_0)$ of §2 which satisfy the following conditions:

(i) $0 < \beta_\nu - \alpha_\nu < \dfrac{\pi}{k_1} \qquad (\nu = 1, 2)$;

(ii) $\mathcal{U}_1(r) \cap \mathcal{U}_2(r) \neq \emptyset$;

(iii) there exist a positive number r_1 and two functions $\vec{f}_1(x)$ and $\vec{f}_2(x)$ such that

  (a) for each ν, the function $\vec{f}_\nu$ is holomorphic in $\mathcal{U}_\nu(r_1)$;

  (b) for each ν, we have $\lim \vec{f}_\nu(x) = \vec{0}$ as $x \to 0$ in $\mathcal{U}_\nu(r_1)$;

  (c) we have

$$\left| \vec{f}_2(x) - \vec{f}_1(x) \right| \leq K \, e^{-\epsilon_1 |x|^{-k}} \qquad \text{in} \;\; \mathcal{U}_1(r_1) \cap \mathcal{U}_2(r_1) \;\; (5.1)$$

  for some positive numbers K and ϵ_1 ;

  (d) for each ν, $\vec{f}_\nu$ is a solution of differential equation (2.1) in $\mathcal{U}_\nu(r_1)$, i.e.

$$x \frac{d\vec{f}_\nu(x)}{dx} \; = \; \vec{G}_0(x) + [\, \lambda(x) + A_0 \,] \, \vec{f}_\nu(x) + x^\mu \, \vec{G}(x, \vec{f}_\nu(x))$$

$$\text{in} \;\; \mathcal{U}_\nu(r_1). \qquad (5.2)$$

We want to prove the following basic lemma.

Lemma 5.1. *We can modify the two functions f_1 and f_2 by some quantities of $O\left(e^{-\epsilon|x|^{-k}}\right)$, where ϵ is some positive constant, so that these modified functions also satisfy conditions (a),(b) and (d) of (iii) given above and that moreover they satisfy the following condition (c′):*

(c′) *we have*

$$\left| \vec{f}_1(x) - \vec{f}_2(x) \right| \leq K' \, e^{-\epsilon_2 |x|^{-k'}} \qquad \text{in} \;\; \mathcal{U}_1(r_2) \cap \mathcal{U}_2(r_2) \;\; (5.3)$$

  for some positive numbers K', r_2 and ϵ_2

Proof:

Step 1: For each $\nu = 1, 2$, changing differential equation (2.1) by

$$\vec{y} = \vec{f}_\nu(x) + \vec{u}_\nu , \qquad\qquad (5.4)$$

we derive another differential equation:

$$x\frac{d\vec{u}_\nu}{dx} = [\,\lambda(x) + A_0\,]\vec{u}_\nu + x^\mu \vec{F}_\nu(x, \vec{u}_\nu) \qquad \text{in } \mathcal{U}_\nu(r_1), \qquad (5.5.\nu)$$

where

$$\vec{F}_\nu(x, \vec{u}) = \vec{G}(x, \vec{f}_\nu(x) + \vec{u}) - \vec{G}(x, \vec{f}_\nu(x)) = \sum_{|p|\geq 1} \vec{u}^p \vec{F}_{\nu,p}(x)$$

is a power series which is uniformly convergent in the domain

$$x \in \mathcal{U}_\nu(r_1), \qquad |\vec{u}| < 2\rho$$

for some positive number ρ and the coefficients $\vec{F}_{\nu,p}(x)$ are holomorphic and bounded in $\mathcal{U}_\nu(r_1)$.

Step 2: We shall apply Lemma 3.1 to each of two differential equations (5.5.ν). To do this, let us assume that

$$\begin{cases} \mathcal{J}_k(1) = \{\, j \in \mathcal{R}_k; \ \Re[\Lambda_j(x)] < 0 \ \text{ in } \mathcal{U}_1(r_1)\} = \{1, \cdots, n_1\} , \\ \mathcal{J}_k(2) = \{\, j \in \mathcal{R}_k; \ \Re[\Lambda_j(x)] < 0 \ \text{ in } \mathcal{U}_2(r_1)\} = \{n_0 + 1, \cdots, n_0 + n_2\} . \end{cases} \qquad (5.6)$$

Then we can construct two differential equations:

$$\begin{cases} x\dfrac{d\vec{w}_1}{dx} = [\,\lambda(x) + A_0\,]_1 \vec{w}_1 + x^\mu \displaystyle\sum_{|p|\geq 2} \vec{w}_1^p \vec{\alpha}_{1,p}(x) , \\[2mm] x\dfrac{d\vec{w}_2}{dx} = [\,\lambda(x) + A_0\,]_2 \vec{w}_2 + x^\mu \displaystyle\sum_{|p|\geq 2} \vec{w}_2^p \vec{\alpha}_{2,p}(x) , \end{cases} \qquad (5.7)$$

where

(i) $\vec{w}_1 = \begin{bmatrix} w_{1,1} \\ \cdots \\ w_{1,n_1} \end{bmatrix}$, and $\vec{w}_2 = \begin{bmatrix} w_{2,n_0+1} \\ \cdots \\ w_{2,n_0+n_2} \end{bmatrix}$;

(ii) the two matrices $[\,\lambda(x) + A_0\,]_1$ and $[\,\lambda(x) + A_0\,]_2$ are the same as in §4;

(iii) if we set

$$\vec{\alpha}_{1,\wp}(x) = \begin{bmatrix} \alpha_{1,\wp,1}(x) \\ \alpha_{1,\wp,2}(x) \\ \cdots \\ \alpha_{1,\wp,n_1}(x) \end{bmatrix}, \qquad \vec{\alpha}_{2,\wp}(x) = \begin{bmatrix} \alpha_{2,\wp,n_0+1}(x) \\ \alpha_{2,\wp,n_0+2}(x) \\ \cdots \\ \alpha_{2,\wp,n_0+n_2}(x) \end{bmatrix},$$

then

$$\begin{cases} \alpha_{1,\wp_1,j}(x) = 0 & \text{if } \lambda_j \neq \sum_{1 \leq \ell \leq n_1} p_{1,\ell}\lambda_\ell , \\[4mm] \alpha_{2,\wp_2,n_0+j}(x) = 0 & \text{if } \lambda_{n_0+j} \neq \sum_{1 \leq \ell \leq n_2} p_{2,n_0+\ell}\lambda_{n_0+\ell} , \end{cases} \tag{5.8}$$

where $\wp_1 = (p_{1,1}, \cdots, p_{1,n_1})$ and $\wp_2 = (p_{2,n_0+1}, \cdots, p_{2,n_0+n_2})$. Let

$$\vec{\phi}_1(x; \vec{c}_1) = \begin{bmatrix} \phi_1(x; \vec{c}_1) \\ \phi_2(x; \vec{c}_1) \\ \cdots \\ \phi_{n_1}(x; \vec{c}_1) \end{bmatrix}, \qquad \vec{\phi}_2(x; \vec{c}_2) = \begin{bmatrix} \tilde{\phi}_{n_0+1}(x; \vec{c}_2) \\ \tilde{\phi}_{n_0+2}(x; \vec{c}_2) \\ \cdots \\ \tilde{\phi}_{n_0+n_2}(x; \vec{c}_2) \end{bmatrix}$$

be general solutions of differential equations (5.7) respectively as constructed in §3.

The most important meaning of differential equations (5.7) is that there exist two formal power series

$$\vec{P}_\nu(x, \vec{w}_\nu) = P_{\nu,0}(x)\vec{w}_\nu + x^\mu \sum_{|\wp| \geq 2} \vec{w}_\nu^\wp \vec{P}_{\nu,\wp}(x) \tag{5.9}$$

such that

(1) $\vec{P}_{\nu,\wp}(x)$ are n-vectors and the $P_{\nu,0}(x)$ are $n \times n_\nu$ matrices respectively;

(2) the entries of $\vec{P}_{\nu,\wp}(x)$ and $P_{\nu,0}(x)$ are holomorphic and bounded in $\mathcal{U}_\nu(r_1)$ respectively;

(3) moreover the quantities

$$|x|^{-\mu'} |P_{\nu,0}(x) - C_\nu|$$

are bounded in $\mathcal{U}_\nu(r_1)$ respectively for some positive number μ' such that

$$\mu' + \Re[\mu_j - \mu_h] > 0 \qquad (\text{for } j, h = 1, \cdots, n),$$

where

$$C_1 = \begin{bmatrix} I_{n_1} \\ O \end{bmatrix}, \qquad C_2 = \begin{bmatrix} O \\ I_{n_2} \\ O \end{bmatrix};$$

here I_{n_ν} is the $n_\nu \times n_\nu$ identity matrix and the O are the zero matrices of suitable sizes;

(4) if we set

$$\vec{P}_{\nu,p}(x) = \begin{bmatrix} P_{\nu,p,1}(x) \\ \cdots \\ P_{\nu,p,n}(x) \end{bmatrix},$$

then we have

$$\begin{cases} P_{1,p,j}(x) = 0 & \text{if } \lambda_j = \sum_{1 \le \ell \le n_1} p_{1,\ell} \lambda_\ell , \\[4mm] P_{2,p,n_0+j}(x) = 0 & \text{if } \lambda_{n_0+j} = \sum_{1 \le \ell \le n_2} p_{2,n_0+\ell} \lambda_{n_0+\ell} ; \end{cases} \tag{5.10}$$

(5) the formal power series

$$\vec{P}_\nu(x, \vec{\phi}_\nu(x, \vec{c}_\nu)) = P_{\nu,0}(x)\vec{\phi}_\nu(x, \vec{c}_\nu) + x^\mu \sum_{|p| \ge 2} \vec{\phi}_\nu(x, \vec{c}_\nu)^p \vec{P}_{\nu,p}(x) \tag{5.11}$$

are formal solutions of differential equations (5.5.ν) respectively.

Thus we modify the two solutions $\vec{f}_1$ and $\vec{f}_2$ of (2.1) by

$$\vec{\psi}_\nu(x, \vec{\phi}_\nu(x, \vec{c}_\nu)) = \vec{f}_\nu(x) + \vec{P}_\nu(x, \vec{\phi}_\nu(x, \vec{c}_\nu)) , \tag{5.12}$$

respectively.

Step 3: We shall apply Lemma 4.1 to the difference

$$\vec{Y} = \vec{\psi}_2(x, \vec{\phi}_2(x, \vec{c}_2)) - \vec{\psi}_1(x, \vec{\phi}_1(x, \vec{c}_1)) \tag{5.13}$$

in the sector $\mathcal{U}_2(r_1) \cap \mathcal{U}_1(r_1)$. First of all, note that $\vec{Y}$ satisfies a linear homogeneous system:

$$x\frac{d\vec{Y}}{dx} = [\lambda(x) + A_0 + x^\mu A] \vec{Y} , \tag{5.14}$$

where

$$A = A(x, \vec{\phi}_1, \vec{\phi}_2) = \sum_{|p_1|+|p_2| \ge 0} \vec{\phi}_1^{\,p_1} \vec{\phi}_2^{\,p_2} A_{p_1 p_2}(x) \tag{5.15}$$

is a power series in $\vec{\phi}_1$ and $\vec{\phi}_2$ whose coefficients $A_{p_1 p_2}(x)$ are holomorphic and bounded in $\mathcal{U}_2(r_1) \cap \mathcal{U}_1(r_1)$. Furthermore, since the direction d of §2 is not singular on the level k, we have

$$j \in \mathcal{R}_k \quad \text{and} \quad \Re[\Lambda_j(x)] < 0 \quad \text{in } \mathcal{U}_2(r_1) \cap \mathcal{U}_1(r_1)$$

if and only if $j \in \mathcal{J}_k(1) \cup \mathcal{J}_k(2)$.

If we apply Lemma 4.1 to system (5.14), we can construct two $n \times n$ matrices Φ and B (cf. (4.5)) satisfying conditions (4.4) and (4.6). This means that, if we set

$$\vec{Z} = \Phi(x, \vec{\phi}_1, \vec{\phi}_2)\vec{Y} ,$$

then $\vec{Z}$ satisfies the linear homogeneous system

$$x\frac{d\vec{Z}}{dx} = [\,\lambda(x) + A_0 + x^\mu B\,]\vec{Z} . \tag{5.16}$$

Therefore, if we utilize the fundamental matrix $\Psi(x, \vec{c}_1, \vec{c}_2)$ of system (4.7), i.e.

$$x\frac{d\vec{V}}{dx} = -\vec{V}[\,\lambda(x) + A_0 + x^\mu B\,] , \tag{4.7}$$

the vector $\vec{\Gamma}(\vec{c}_1, \vec{c}_2)$ defined by

$$\vec{\Gamma}(\vec{c}_1, \vec{c}_2) = \Psi(x, \vec{c}_1, \vec{c}_2)\vec{Z} = \Psi(x, \vec{c}_1, \vec{c}_2)\Phi(x, \vec{\phi}_1, \vec{\phi}_2)\vec{Y}$$

is independent of x. Furthermore

$$e^{\Lambda(x)}\vec{\Gamma}(\vec{c}_1, \vec{c}_2) = e^{\Lambda(x)}\Psi(x, \vec{c}_1, \vec{c}_2)\Phi(x, \vec{\phi}_1, \vec{\phi}_2)\vec{Y} \tag{5.17}$$

can be written in the following form:

$$e^{\Lambda(x)}\vec{\Gamma}(\vec{c}_1, \vec{c}_2) = \sum_{|p_1|+|p_2|\geq 0} \left(e^{[\Lambda]_1}\vec{c}_1\right)^{p_1} \left(e^{[\Lambda]_2}\vec{c}_2\right)^{p_2} \vec{\Gamma}_{p_1 p_2}(x) \tag{5.18}$$

where

$$\begin{cases} \Lambda(x) = diag\,(\Lambda_1(x), \cdots, \Lambda_n(x)) , \\ [\Lambda(x)]_1 = diag\,(\Lambda_1(x), \cdots, \Lambda_{n_1}(x)) , \\ [\Lambda(x)]_2 = diag\,(\Lambda_{n_0+1}(x), \cdots, \Lambda_{n_0+n_2}(x)) , \end{cases}$$

and the $\Gamma_{p_1 p_2}(x)$ are holomorphic in $\mathcal{U}_2(r_1) \cap \mathcal{U}_1(r_1)$; also

$$|\,\Gamma_{p_1 p_2}(x)\,| \leq \gamma_{p_1 p_2}\,|\,x\,|^{\mu_{p_1 p_1}}$$

in $\mathcal{U}_2(r_1) \cap \mathcal{U}_1(r_1)$ for some positive numbers $\gamma_{p_1 p_2}$ and some real numbers $\mu_{p_1 p_2}$.

Remark: If we set $\vec{c}_1 = \vec{0}$ and $\vec{c}_2 = \vec{0}$ in (5.17), we get

$$\vec{\Gamma}(\vec{0}, \vec{0}) = x^{-A_0} e^{-\Lambda(x)} \left[I + x^{\mu'} \Phi_0(x) \right] (\vec{f}_2(x) - \vec{f}_1(x)) . \qquad (5.19)$$

Step 4: Letting x tend to zero in $\mathcal{U}_2(r_1) \cap \mathcal{U}_1(r_1)$ we can compute $\vec{\Gamma}(\vec{c}_1, \vec{c}_2)$. In fact, if we set

$$\vec{\Gamma}(\vec{c}_1, \vec{c}_2) = \begin{bmatrix} \Gamma_1(\vec{c}_1, \vec{c}_2) \\ \Gamma_2(\vec{c}_1, \vec{c}_2) \\ \cdots \\ \Gamma_n(\vec{c}_1, \vec{c}_2) \end{bmatrix} ,$$

we have

$$\Gamma_j(\vec{c}_1, \vec{c}_2) = 0 \quad \text{if} \quad \tau_j < k \text{ or } \tau_j = k \text{ and } j \notin \mathcal{J}_k(2) \cup \mathcal{J}_k(1) .$$

Let us look at Γ_j for j such that $\tau_j = k$ and that $j \in \mathcal{J}_k(2) \cup \mathcal{J}_k(1)$. For such j, we have

$$\Gamma_j(\vec{c}_1, \vec{c}_2) = \sum_{|p_1| + |p_2| \geq 0} \vec{c}_1^{p_1} \vec{c}_2^{p_2} \Gamma_{j, p_1 p_2} \qquad (5.20)$$

where $\Gamma_{j, p_1 p_2} = 0$ if

$$\Re \left[\Lambda_j(x) - \sum_{1 \leq \ell \leq n_1} p_{1,\ell} \Lambda_\ell(x) - \sum_{1 \leq \ell \leq n_2} p_{2, n_0 + \ell} \Lambda_{n_0 + \ell}(x) \right] > 0 \qquad (5.21)$$

in $\mathcal{U}_2(r_1) \cap \mathcal{U}_1(r_1)$. This means that the right-hand side of (5.20) is a polynomial in $\vec{c}_1$ and $\vec{c}_2$.

Step 5: More precisely speaking, for $j \in \mathcal{J}_k(2) \cup \mathcal{J}_k(1)$ we can derive

$$\Gamma_j(\vec{c}_1, \vec{c}_2) = \begin{cases} \xi_j + c_{2,j} + \cdots & \text{if } j \in \mathcal{J}_k(2) \text{ and } j \notin \mathcal{J}_k(1) , \\ \xi_j + c_{2,j} - c_{1,j} + \cdots & \text{if } j \in \mathcal{J}_k(2) \cap \mathcal{J}_k(1) , \\ \xi_j - c_{1,j} + \cdots & \text{if } j \in \mathcal{J}_k(1) \text{ and } j \notin \mathcal{J}_k(2) , \end{cases}$$
$$(5.22)$$

where $\xi_j = \Gamma_j(\vec{0}, \vec{0})$ and $\cdots$ denotes terms containing only those $c_{\nu,\ell}$ such that $\Re[\Lambda_j(x)] < \Re[\Lambda_\ell(x)]$ in $\mathcal{U}_2(r_1) \cap \mathcal{U}_1(r_1)$. Therefore we can fix arbitrary constants $\vec{c}_1, \vec{c}_2$ in such a way that

$$\Gamma_j(\vec{c}_1, \vec{c}_2) = 0 \quad \text{for} \quad j \in \mathcal{J}_k(2) \cup \mathcal{J}_k(1) .$$

This completes the proof of Lemma 5.1.

Remark: The argument given above is completely formal. However, by utilizing those results obtained in M.Hukuhara[5], M.Iwano[6] and J.-P.Ramis and Y.Sibuya[10], we can justify our computations analytically without any difficulties.

6.   Proof of Theorem 2.1

We shall utilize the same notations as in §2,   and apply Lemma 5.1 to $\{\mathcal{U}_\nu(r_1), \mathcal{U}_{\nu-1}(r_1)\}$ and $\{\vec{f}_\nu(x), \vec{f}_{\nu-1}(x)\}$ for each ν.

To do this, set

$$\mathcal{J}_k(\nu) = \{\, j \in \mathcal{R}_k;\, \Re[\Lambda_j(x)] < 0 \quad \text{in } \mathcal{U}_\nu(r_1)\} = \{\, j_{\nu,1}, \cdots, j_{\nu,n_\nu}\,\}, \quad (6.1)$$

and

$$\vec{c}_\nu = \begin{bmatrix} c_{\nu,j_{\nu,1}} \\ c_{\nu,j_{\nu,2}} \\ \cdots \\ c_{\nu,j_{\nu,n_\nu}} \end{bmatrix}. \tag{6.2}$$

Then for each ν, we have the following system of equations:

$$
\begin{aligned}
0 &= \Gamma_j(\vec{c}_{\nu-1}, \vec{c}_\nu) \\
&= \begin{cases} \xi_j + c_{\nu,j} + \cdots & \text{if } j \in \mathcal{J}_k(\nu) \text{ and } j \notin \mathcal{J}_k(\nu-1), \\ \xi_j + c_{\nu,j} - c_{\nu-1,j} + \cdots & \text{if } j \in \mathcal{J}_k(\nu) \cap \mathcal{J}_k(\nu-1), \\ \xi_j - c_{\nu-1,j} + \cdots & \text{if } j \in \mathcal{J}_k(\nu-1) \text{ and } j \notin \mathcal{J}_k(\nu), \end{cases}
\end{aligned}
$$
$$(6.3.\nu)$$

where $\xi_j = \Gamma_j(\vec{0}, \vec{0})$ and $\cdots$ denotes terms containing only those $c_{\nu,\ell}$ and $c_{\nu-1,\ell}$ such that $\Re[\Lambda_j(x)] < \Re[\Lambda_\ell(x)]$ in $\mathcal{U}_\nu(r_1) \cap \mathcal{U}_{\nu-1}(r_1)$ (cf. (5.22)).

Let us classify those j in $\mathcal{R}_k$ according to Case A and Case B of figure 1 of §2. Then, since the direction d is not singular on the level k, we can choose, for each ν, a point x_ν in $\mathcal{U}_\nu(r_1) \cap \mathcal{U}_{\nu-1}(r_1)$ in such a way that

$$\Re[\Lambda_j(x_\nu)] \begin{cases} < \Re[\Lambda_j(x_{\nu+1})] & \text{if } j \text{ is in Case A,} \\ > \Re[\Lambda_j(x_{\nu+1})] & \text{if } j \text{ is in Case B} \end{cases} \tag{6.4}$$

(cf.   J.-P.Ramis and Y.Sibuya[10]). Let us order all of equations $(6.3.\nu)$ $(\nu = 2, 3, \cdots, N)$ by the order of $\{\, \Re[\Lambda_j(x_\nu)]\, ; j \in \mathcal{J}_k(\nu), \quad \nu = 2, 3, \cdots, N\,\}$. Then we can solve those equations successively. Thus we can complete the proof of Theorem 2.1.

7.   Acknowledgements

The main part of this paper was prepared at the University of Groningen, the Netherlands, in a series of lectures in seminars, while the author was a guest researcher supported partially by the Netherlands Organisation for Scientific Research and the Department of Mathematics, University of Groningen. The author wishes to thank these institutes for the invitation and their hospitality.

He also expresses his appreciation to Professor B.L.J.Braaksma for many useful discussions of multisummability and the Stokes phenomena which are the background of this paper. The author is also partially supported by a grant from the National Science Foundation.

8. References

1. W. Balser, B. L. J. Braaksma, J-P. Ramis and Y. Sibuya, Multisumma-bility of formal power series solutions of linear ordinary differential equations, Asymptotic Analysis 5(1991) 27-45.

2. B. L. J. Braaksma, Multisummability and Stokes multipliers of linear meromorphic differential equations, J. Differential Equations 92(1991) 45-75.

3. B. L. J. Braaksma, Multisummability of formal power series solutions of nonlinear meromorphic differential equations, Ann. Inst. Fourier, Grenoble, 42(1992), to appear.

4. J. Ecalle, Calcul accélératoire et applications, Act. Math., Ed. Hermann, Paris, to appear.

5. M. Hukuhara, Intégration formelle d'un système d'équations différentielles non linéaires dans le voisinage d'un point singulier, Ann. di Mat. Pura Appl., 19(1940) 34-44.

6. M. Iwano, Intégration analytique d'un système d'équations différentielles non linéaires dans le voisinage d'un point singulier,I, II, Ann. di Mat. Pura Appl., 44(1957) 261-292, 47(1959) 91-150.

7. B. Malgrange and J-P. Ramis, Fonctions multisommables, Ann. Inst. Fourier, Grenoble, 42(1992) 353-368.

8. J. Martinet and J-P. Ramis, Elementary acceleration and mutisumma-bility, Ann. Inst. Henri Poincaré, Physique Théorique, 54(1991) 331-401.

9. J-P. Ramis, Les séries k-sommables et leurs applications, Analysis, Microlocal Calculus and Relativistic Quantum Theory, Proc. "Les Houches" 1979, Springer Lecture Notes in Physics 126(1980) 178-199.

10. J-P. Ramis and Y. Sibuya, Hukuhara's domains and fundamental existence and uniqueness theorems for asymptotic solutions of Gevrey type, Asymptotic Analysis 2(1989) 39-94.

Printed in the United States
By Bookmasters